advance praise for **THE POWER OF WHY**

"A truly great book changes the way you think. It gives you a new perspective, one that leads to positive results. This is a truly great book."
—David Chilton, author of *The Wealthy Barber* and *The Wealthy Barber Returns*

"Packed with insights, analysis, and examples drawn from a wide range of individuals and companies, *The Power of Why* will convince readers of the profound value of the spirit of curiosity."
—Gretchen Rubin, author of *The Happiness Project*

"There are books filled with anecdotes about innovation or you can read academic theories about innovation. Neither get you very far. What Amanda Lang has done is offer interesting stories that are linked together in a way that teaches us the theories and just how to innovate."
—Clayton M. Christensen, Harvard Business School professor and author of *How Will You Measure Your Life?*

"The ability of a company to innovate is critical for survival let alone growth, and *The Power of Why* provides great insights into how individuals can transform organizations and industries. A great read on how to find success."
—Gordon M. Nixon, President and CEO, RBC

"It is standard fare for political and economic commentators to bemoan Canada's lack of innovation and productivity. It is also standard fare that they rarely provide answers beyond their statement of concern. Amanda Lang has done much more than that. Her insight into what sparks curiosity and how it is the foundation of innovation and improved productivity should give pause for thought to all of those whose interests run the range from early learning to venture capital. *The Power of Why* shows us a way to a higher quality of life and to a stronger economy."
—The Right Honourable Paul Martin

"This is a lot more than a business book, it's a life book. Just pick it up for ten minutes and you'll find yourself thinking in some exciting, new, and yes, innovative ways."
—Peter Mansbridge, chief correspondent for CBC News

"*The Power of Why* is a compelling and superbly written book showing how Canada can and must support creativity, starting in our schools, to fuel innovation and positive change."
—David Miller, former mayor of Toronto

"A wonderful book, *The Power of Why* demystifies the tricky subject of innovation. It is a must-read for all innovation policy types and every budding innovator."
—Roger Martin, Dean, Rotman School of Management, University of Toronto

THE POWER OF WHY

The Power of Why

AMANDA LANG

Collins

The Power of Why

Published by Collins, an imprint of HarperCollins Publishers Ltd

First published by Collins in a hardcover edition: 2012
This trade paperback edition: 2014

HarperCollins books may be purchased for educational, business, or sales promotional use through our Special Markets Department.

HarperCollins Publishers Ltd
2 Bloor Street East, 20th Floor
Toronto, Ontario, Canada
M4W 1A8

www.harpercollins.ca

Library and Archives Canada Cataloguing in Publication information is available upon request

ISBN 978-1-44341-319-0

Printed and bound in the United States
RRD 10 9 8 7 6

For Julian and Madeleine,
the best innovations of all

CONTENTS

THE POWER OF WHY

INTRODUCTION

Standing in the cavernous home workshop he'd built with his own hands, Steve Gass took a deep breath and steeled himself to do the unthinkable: turn on his table saw and try to touch the whirring blade, to see whether it chopped off his finger. If the safety mechanism he'd toiled over for months actually worked, his finger would be just fine. If not, well . . . Gass, who is right-handed, had decided he'd use the ring finger of his left hand. Just in case.

Gass, who lives in Portland, Oregon, believed in the new and improved table saw he'd devised, which was outfitted with a sensor that could, theoretically anyway, detect human flesh and, in a few thousandths of a second, automatically stop the blade. He was so sure it was a game-changing innovation that he'd given up his lucrative patent law practice to focus on building a prototype. And the prototype worked on hot dogs; he knew that already, having held many a sausage up against the blade, which stopped pretty much instantly, leaving only a minor nick. So the blade *should* stop when it sensed a finger coming, too. Yet . . .

There he stood, willing his brain to forget that the blade was moving 100 miles an hour or more, so much faster than humans can react that "oftentimes, you cut through three fingers before you even flinch. People just misplace their hand on a piece of a wood and wind up feeding their fingers through the blade," says Gass, a lean, fit man who doesn't have a lot of hair left and wasn't eager to part with a digit, too. His heart was beating crazily and the muscles in his left arm were cramping from the tension. "Every fibre of your being says this is not something you should be doing."

It was a lesson he'd learned the hard way at the age of four, when, fooling with his father's tools, he'd managed to lop off a chunk of his thumb. While he recalls with great clarity the details leading up to the accident—his small hand pretending to work a joiner the way his dad did, holding the switch halfway on to allow it to move—his brain has mercifully blocked out what followed, fast-forwarding past the tears, the blood and the trip to the emergency room for stitches while his parents thanked God it hadn't been worse. But even though he couldn't remember all those bits, they were surely lurking somewhere in the recesses of his memory, triggering some primal, and frankly quite rational, fears at this particular moment.

However, he still needed to prove that his innovation actually worked. He knew that when SawStop, as he'd named his safe table saw, was unveiled for the first time at a trade show, people would watch the hot dog demonstration politely, then ask, "But how do you know it will stop for a *finger*?"

Good question. And for someone like Gass, a question is like an itch. He can't ignore it or will it away. If he tries not to think about it, it only intrudes more insistently; relief comes only when he finds an answer. His journey to this moment, in fact, started with a question. About six months earlier, puttering around in his workshop, he'd happened to glance over at his table saw and a question popped into his head: Hey, I wonder if there's a way to stop the blade fast enough to prevent injury? He wasn't looking to quit his job or make money, he was just curious, intellectually, and in the habit of following his curiosity where it led him. He'd always been like that. As a kid, he'd been keen on taking things apart to see how they worked, and by doing that, had learned to fix just about anything (his parents had pretty much insisted on the fixing part).

And so it was no surprise to anyone who knew him when, in the spring of 2000, after a few unpleasant minutes of hesitation standing in front of the SawStop and several false starts, Steve Gass jabbed his finger right into the teeth of the whirling saw blade. He didn't feel he had a choice. After all, he had a question and needed to find the answer. For Gass, curiosity is what gives him purpose.

INNOVATION IS SIMPLER THAN YOU THINK

In industries where fast-paced change is the norm, innovation has become the holy grail. Companies that are scrambling to stay relevant hold "blue sky" brainstorming sessions and bring in wildly expensive consultants and stick beanbag chairs in the boardroom because someone read somewhere that that's how they do things in Silicon Valley.

One of the reasons there's such a big focus on window dressing and expert intervention is that many people believe innovation is hugely difficult. In fact, a lot of people believe that the ability to innovate and create something new is one you're born with—and if not, tough luck, you'll never be able to do it. According to a recent survey conducted by the European Centre for Strategic Innovation, "68% of business leaders firmly believe that great innovators are born and cannot be made."

However, scientists have shown that the exact opposite is true. For instance, a landmark study of identical twins who were separated at birth found that although 80 percent of the variation on IQ tests is attributable to genetics, only 30 percent of performance on creativity tests can be explained that way. In other words, 70 percent of creativity is related to environment, which means that it's entirely possible for just about anyone to learn to think more innovatively.

Now, this doesn't mean that everyone has an equal shot at being the next Steve Jobs. Most social scientists differentiate between two distinct types of creativity: big C—the kind of inventive genius that Jobs, Mozart and da Vinci had—and small c, the more common variety of innovative creativity that a session musician or a good surgeon or, for that matter, Steve Gass might have. Small-c creativity is the 70 percent sort that's not genetic and depends more heavily on environment, attitude, mindset and exposure to enrichment rather than on one-in-a-million ability. Recently, some researchers have suggested a third variety: mini c, the kind that people demonstrate when they're concocting a recipe or solving a math problem.

Creativity doesn't necessarily mean you'll be innovative. Innovation is all about improvement, changes that make things better; a politician whose main gift is fabricating lies is creative but not innovative. But, generally, the ability to think creatively is necessary for innovation, and here's the good news: while it's nice to have big-C genius or even small-c expertise, the truth is that innovation really only requires mini-c aptitude.

Innovation is not, as many people believe, synonymous with invention and therefore out of the reach of the average person. It's simpler than you might think. You don't have to create something mind-blowing and entirely new, like the automobile or the Internet. Often, innovation simply means making incremental improvements to something that already exists, like a table saw. And frequently that's accomplished by borrowing and adapting an idea or approach or technology from another field altogether; Steve Gass, for instance, adapted the flesh-detecting sensor used on touch screens and installed it on his saw to serve a different purpose. Old plus old does, sometimes, equal new.

Other times, the idea behind an innovation is so basic that it's hard to imagine it never occurred to anyone before. It looks less like innovation than it does common sense. Consider this one: reward loyalty. Simple concept, yet it was a big innovation when airlines introduced frequent flyer miles. It changed the way a lot of people travel.

Another great example: Federal Express. The company didn't invent a new product. Its innovation was simply to introduce the idea of reliable overnight delivery. It used to be that when you shipped a parcel out of town, it passed through multiple hands. From your office, it might go to the company mailroom, where a clerk would arrange to have it sent over to the post office, and the post office would have it sent over to the airport, where it was loaded on a plane, and the plane would fly it to another airport, where someone would unload it and deliver it to another post office, which would then be in charge of getting it to its destination, where another mailroom clerk would deliver it to the recipient. At every point in that chain there is the possibility for error, and for your package to be delayed or lost. And if that happens, who is held accountable? What FedEx innovated was simply to own the whole chain. It didn't invent a single thing—not the idea of postal delivery, or the vehicles used to accomplish it. All it innovated was the idea of reliable speed in mail delivery, and that idea turned out to be worth a whole lot of money.

Innovation usually starts pretty much the way it did with Steve Gass. He noticed a problem—table saws are really dangerous—and asked himself if there was some way to fix it. "I was just wondering, 'Could this be done?' That was the 'watershed' thought process," he explains wryly.

Virtually all innovation, whether it involves a different way of cooking a turkey or a new kind of online service, starts with a question like Gass's. A simple one, such as, Is there a way to do this better/cheaper/faster?

The difference between a dreamer and an innovator is that the innovator doesn't stop there—the questions just keep on coming. Gass, for instance, wondered how deep a cut from a saw blade would be acceptable (he settled on one-eighth of an inch; the finger might bleed and even need stitches, but it wouldn't be amputated). How quickly would the blade need to stop? How much force would be necessary to stop it? How could you be sure the saw was just as good at cutting wood as a regular one? What kind of sensor could differentiate between wood and human flesh? How could you ensure it was reliable?

Sometimes, along the way, he got the answers wrong, but that didn't discourage him. "A big part of innovation is being a critic, and knowing what you should discard," Gass observes. "It's crucial to ask, 'Why won't this work? What's wrong with it?'"

A lot of us, though, just don't do this. On the job and at home, many of us hit on an answer that sounds "right," or that others approve of, then figure question period is over. We burrow down and focus on implementing whatever solution we've devised, unwilling to revisit or rethink it. And that's a mistake. Sometimes it's the follow-up question, or the one after that, that is going to yield the game-changing revelation. And sometimes getting there will require 100 more questions. Just stopping at the first plausible response is how a lot of us get stuck and find ourselves unable to solve problems, both at work and at home. The rush to get the questions over with and land on an answer is also why we can wake up one day and realize that we're trapped in the wrong line of work, the wrong relationships—the wrong lives, even. We just didn't ask the right questions, or enough questions, and after seizing on a conclusion, didn't make a regular practice of critically re-evaluating it in light of our actual experiences to see if it really *was* the right answer.

The willingness to keep asking is much more important to successful innovation than big-C originality is. Strangely, however, many of us simply don't do that.

But it's not that we can't. It's that we've forgotten how.

WHAT INNOVATORS (AND TODDLERS) KNOW

As a business journalist, I've had the opportunity to interview a lot of people like Gass who've come up with a new product or process or service, or who've found a new way to run an organization. For the most part, they're impressive, interesting people (with the occasional egomaniac thrown in for good measure). But they're rarely big-C, dazzling geniuses. The main difference between them and the rest of us is that they ask more and better questions, and they are more driven to find answers and embrace them, even if the answers are at first not what they wanted or expected to find. They have less in common with Einstein, frankly, than with young children.

Their minds are open to new possibilities in much the same way my son's mind was when he turned three and began asking questions. Non-stop. Where do goldfish go after they die? Why do I have to go to bed and you don't? What will happen if I pour my milk into my spaghetti? How do I make this train work?

Even babies and toddlers form hypotheses and then gather evidence to test them in much the same way scientists do, as psychologist Alison Gopnik has pointed out. Little kids perform one experiment after another: Hey, this key ring looks tasty, I'm putting it in my mouth. Oh. Gross. Maybe if I put it in the toilet something exciting will happen. How do I flush this thing? Oh, press on that silver thingy! Here comes Dad, and he's yelling. This *is* exciting! If I stick my head in there, will he yell even louder?

What propels kids' experimentation is curiosity. Kids are all about ignoring conventional wisdom and finding out for themselves. Sometimes, admittedly, the results aren't pretty. Remember that impulse to lick the frozen metal post because you'd heard your tongue might stick?

That same sort of impulse, coupled with adult analytic reasoning, was driving Steve Gass when he touched the blade of the SawStop, which, I'm happy to report, did indeed stop. "It was a little anticlimactic actually," Gass says, "just a tiny scratch on my finger." (And a good thing, too, since Gass, the kind of guy who enjoys going over waterfalls in his kayak, is nevertheless a bit of a wuss when it comes to blood. "I tend to get a little faint" is how he puts it.)

Why aren't more of us as curious as he is? Why, when it's so easy and natural for little kids to question and challenge and test *everything*, have so many adults lost these habits? Why do we equate "childlike wonder" with naïveté, when it's clearly linked to success in ways that are tangible and quantifiable in the world of business? Is it possible to retrain ourselves and reignite our natural curiosity? How, exactly?

These are the questions I'll be answering in the pages that follow, through the stories of innovative individuals and companies. You'll see how curiosity drives progress—

and not just in the business world. As I learned by talking to innovators and researchers, it's entirely possible to adapt some of the approaches and techniques that work so well in the corporate world and use them to solve everyday problems in your personal life, too. The best part is that you don't need an MBA or a sophisticated grasp of economic theory. You don't even need to know how to read a spreadsheet. All you really need is the courage to ask questions and to embrace the answers, whether or not they're to your liking.

And the benefits are huge. Whether you're a CEO looking to grow your business or a frazzled parent hoping to improve your relationship with your kids, asking the right questions, in the right way, at the right time greatly increases your chances of thinking innovatively so that you can solve problems and spot new possibilities. Curiosity is directly linked to success—however you measure it. But curiosity doesn't just have utilitarian value. It doesn't just help you find solutions and make progress and understand yourself and the rest of the world better. It can actually help you have a better life, one in which you're engaged, energized, fulfilled and constantly learning.

That, ultimately, is the power of why.

CHAPTER ONE

What Happens to Curiosity?

We are born curious. Thank goodness. If babies didn't have an innate drive to figure out how the world works, they wouldn't learn very much. Curiosity keeps them interested, alert, observant and focused. Later in life we learn to look like we're paying attention, out of politeness, or we pay attention merely to pass the test or because we suspect the information may be helpful in the future. But babies and toddlers? They haven't learned to fake it. If they're not interested, they simply zone out or cry or experiment with something they are curious about, like the physics of flinging a bowl of applesauce across the room, or finding out what's in that cupboard with the child lock.

But very early on, and often unwittingly, we begin to train curiosity out of kids. Think of the messaging: curiosity killed the cat, led Little Red Riding Hood off the straight and narrow path and didn't work out so well for Pandora, either. Think of the warnings about talking to strangers. Think about Eve, for goodness sake, who basically got booted out of the Garden of Eden because she wondered what apples taste like. When they are still very little, children begin to receive contradictory messages about curiosity. Asking whether *c-a-t* spells *cat* is good. Asking "Why is that guy so fat?" is impolite. Asking your great-aunt whether she'd like another biscuit is nice. Being curious about and open to strangers is dangerous. Asking whether it's all right to go in the pool is sensible. Challenging authority and tradition is disrespectful.

Admonitions about the dangers of curiosity usually kick in at about the same time that children start actively searching for causal explanations, seeking information that can help them predict and interpret events and figure out the world. Or put another way, one that's familiar to anyone who has spent any time at all with a toddler: they ask questions. A *lot* of questions—dozens per preschooler per hour, according to researchers. By the age of three, quite a few of these questions start with the word *why*. For adults, this can quickly get irritating, especially when your best explanations just elicit yet another *why*. Many adults view a never-ending stream of questions as an attention-seeking gambit and brush off kids' questions or ignore them entirely or bark, "Because I said so, that's why!" Researchers report that almost 40 percent of the time, either adults simply don't respond to young children's questions or their response is some variation of "Get lost." (You might wonder why so many adults respond this way; maybe it's because it's how they remember being treated as children.)

But according to a recent study that closely tracked what children do *after* asking questions, they are not seeking attention. Information really is what they're after. "When preschool children ask 'why?' questions, they are not merely trying to prolong the conversation (as previously suspected by many parents and researchers alike). Upon receiving an explanation, children often end their questioning and react with satisfaction," the researchers reported. It's when kids *don't* get the explanatory information they're seeking that the endless *why*s start. The reason: a thirst for knowledge, not an uncanny talent for annoying grown-ups.

However, if questions don't get answered or are actually rebuffed, many kids simply conclude that there's no point in asking. That's exactly what we don't want them to do, for a number of reasons. For starters, highly curious kids learn more; the more they find out, the more they realize they don't know and the deeper they dig for information, whether the topic they're interested in is computers or rap or chemistry. Curiosity is, therefore, strongly correlated with intelligence. For instance, one longitudinal study of 1,795 kids measured intelligence and curiosity when they were three years old, and then again eight years later. Researchers found that kids who had been equally intelligent at age three were, at eleven, no longer equal. The ones who'd been more curious at three were now also more intelligent, which isn't terribly surprising when you consider how curiosity drives the acquisition of knowledge. The more interested and alert and engaged you are, the more you're likely to learn and retain. In fact, highly curious kids scored a full twelve points higher on IQ tests than less curious kids did.

Furthermore, curiosity is intrinsically rewarding. If you've ever watched little kids absorbed in trying to figure out how to play a new game or solve a puzzle, you know what I mean. The desire to acquire more IQ points isn't what motivates them. What's driving them is more self-interested: pleasure. It *feels* good to be interested, to be driven to explore and find out new things. Sometimes it feels risky or even aggravating not knowing what the answer is or what will happen next. But always there's a sense of mental alertness. And that sure feels better than being bored and disengaged.

Curious kids learn *how* to learn, and how to enjoy it—and that, more than any specific body of knowledge, is what they will need to have in the future. The world is changing so rapidly that by the time a student graduates from university, everything he or she learned may already be headed toward obsolescence. The main thing that student needs to know is not *what* to think but *how* to think in order to face new challenges and solve new problems.

"In the industrial economy, the person who wins is the expert," explains Claude Legrand, co-author of *Innovative Intelligence.* "In the knowledge economy, the person who wins is the one who has the process to solve complex problems." That's because, in the knowledge economy, the goalposts are shifting constantly. What's hot today may be old news in six months. Developing the processes to cope with the challenges this will pose in terms of our jobs, Legrand believes, is all about being receptive to change, and possessing the mental and emotional flexibility and desire to continue learning—having a curious mentality, in other words, rather than an expert mentality.

THE MYTH OF EXPERTISE

Many of us are still hanging on to an expert mentality, believing that innovation is connected to expertise. Some of the time this is true. Facebook wouldn't exist if Mark Zuckerberg hadn't been extraordinarily knowledgeable about computers and the Internet. New drugs are developed by highly trained scientists, not high school students.

But oftentimes, innovators have no particular expertise. They simply have a lot of questions. A curious mentality, in other words.

Steve Gass, for instance, was a patent lawyer when he created the SawStop. He wasn't an engineer and had no experience building power tools. He was able to figure out how to jerry-rig a used table saw he bought for $200 because one of his hobbies is building radio-controlled airplanes, but he wouldn't have gone to the trouble if he hadn't been so curious to see whether his idea could really work. He had a busy practice and lots of other interests, like kayaking. He wouldn't have stuck with the saw experiment if he hadn't enjoyed the process of looking for answers.

For him, real-world problems have the same allure that crossword puzzles do for other people. They're fun, and he wants to solve them—so much that he'll stick with a problem as long as it takes to get an answer.

Why is he like this? Partly temperament, probably. He just enjoys challenging standard thinking, and understands this is both a strength and a weakness; some people find him argumentative—"just ask my ex-wife," he says with a laugh. But there's another reason, too. When he was growing up, his parents let him break things in order to learn how to fix them. They encouraged his experiments (though they were pretty distressed when he demagnetized their brand-new colour TV, so that, until the repairman arrived, the images were not only black and white but also distorted). Being allowed to make mistakes, he learned to experience them as part of the learning process rather than as catastrophic events, and he learned how to persevere until he figured something out. By the time he was at university, he had his own tool box, and when he came home for visits, his parents greeted him at the door with a long to-do list: fix the busted washing machine, figure out why the refrigerator door won't close properly . . . To Gass, these were less chores than enjoyable diversions.

His interest in problem solving fuels his tenacity, which is a good thing, as he's needed it with SawStop. His invention seemed like a slamdunk, and the prototype won innovation awards and, in 2002, made the 100 Best New Innovations list in *Popular Science* magazine. It's easy to see why: standard table saws are incredibly dangerous. *Every single day,* they cause eleven amputations and eleven fractures in the United States alone. All told, there are 67,300 "medically-treated blade contact injuries annually," according to the U.S. Consumer Product Safety Commission. The total market for table saws is about $200 million, but the annual cost to the U.S. economy of the injuries they cause is ten times that—the safety commission estimates it's $2.36 billion.

But just as automakers fought the installation of airbags, so did the power tool industry fight—and continues to fight—SawStop. Tool manufacturers have never had to care about fingers, but about profit margins. Licensing Gass's technology would cost them money, but that would be peanuts compared to the capital costs they'd incur overhauling their existing production lines and products. And they'd *have* to do it. If an injury mitigation system crept into the mainstream market, it would be impossible to manufacture a table saw without it because of the liability issues. Speaking of which, if a toolmaker marketed a "safe" saw, who would be liable in the event of an accident? So far, consumers had accepted that liability as 100 percent theirs—would SawStop shift that balance? Toolmakers didn't want to find out. No thanks, they told Gass. Not only that, they seemed determined to undermine SawStop, criticizing its technology.

And so it was that Gass and the two other patent lawyers who'd become his business partners asked themselves a crucial question: Should they try to get their old jobs back, or learn how to manufacture SawStops? They'd never run a business or manufactured a thing in their lives, and they'd need to raise capital. The only certainty was that there would be some very lean years. But the prospect of a steep learning curve was energizing; all three liked figuring out how to do new things, and by 2004, they had.

"We started selling saws in November, and then in March 2005 I got a call from a customer," Gass remembers. "He said, 'Steve, we had a guy run his hand into his saw today.' I'm just waiting, my stomach dropped, and then he said, 'Worked just like you said! He's got a little nick.' It was a huge relief. Two weeks later we got the next call. Now I think we probably get a finger-save a day."

Today, the SawStop isn't just safer than standard saws. "How can we make this better?" is the constant refrain that has driven incremental improvements and the development of new features, such as a better fence to hold back wood chips and particles. Indeed, the entire spirit of the company revolves around challenging the status quo—and each other. "How do you know?" is another constant refrain. Gass says, "We had a guy come out to visit, a former client looking to invest in the company, and as he was leaving he said, 'Gosh, is everything okay here? You guys seem to be going at each other.' I was flabbergasted. From our perspective, questioning and challenging is just normal, there's nothing heated about it."

Meanwhile, SawStop has asked the U.S. Consumer Product Safety Commission to rule on whether injury mitigation systems should be mandatory for table saws—a gamble, as realistically the major manufacturers may just wind up ripping off SawStop's technology and launching an endless patent fight. But, despite the ongoing uncertainty—and the fact that his bank balance is much lower than it would be had he remained in law—Steve Gass has no regrets. "This is my dream job," he says, "because I get to work on interesting technology and it's challenging, and I'm proud that we prevent mutilations and injuries."

He didn't find his dream job because he was an expert. He got where he is because he's naturally curious, and that makes learning new things and taking on new challenges both enjoyable and fulfilling. He takes his measure not through the size of his paycheque (though a large one, he agrees, is pretty nice) but through the volume of what he's learned, and how he's been able to apply it in practical ways to make the world a little better.

The knowledge economy needs people like Steve Gass, people for whom the same old same old isn't good enough, people who like to tinker with things in the hope of coming up with something even better.

CURIOSITY IN THE CLASSROOM

There's been a lot of hand-wringing over the past forty years about the need for schools to promote creativity and original thinking, but there hasn't been widespread recognition that creativity is an *outcome.* It's triggered by curiosity, and in my view, that's what schools should be promoting first and foremost. Let's face it: passive, incurious kids are not creative. They're not churning out imaginative stories or figuring out new stuff in the chemistry lab or fooling around with a computer program to try to make it do something different. Kids who aren't very curious are couch potatoes, intellectually speaking. The reality is that before anyone can do anything innovative or original, there's got to be a sense of wonder or at least a spark of interest, and a whole bunch of questions.

But while schools are terrific at disseminating information, most of them are quite a bit less terrific at promoting curiosity. In many schools, there is, as one researcher aptly put it, "a pedagogy of 'intellectual hide-and-seek' in which teachers hold all the correct answers and students aim to seek out, memorize, and parrot back those answers." Students learn to suppress their own insights and ideas and, instead, try to figure out what the teacher wants to hear. The system is set up so that kids will converge on the one right answer, rather than thinking divergently—coming up with several novel and unexpected possibilities—and asking questions of their own. According to the researcher, "Such practices not only underestimate the importance of imaginative thinking, but deaden the personal value of the information being taught to students."

In an educational system in which productivity is measured by hours logged per task, number of worksheets completed and scores on standardized tests, it doesn't make a whole lot of sense to prompt kids to ask more questions unless the questions are about what's going to be on the test. In many classrooms, stopping to encourage and mull over questions that aren't procedural or directly related to the material at hand is viewed as wasting time. It's no big surprise then that most kids come to school bursting with questions, but exit, a dozen or so years later, asking very few. Curiosity declines from one grade to the next, and the reason isn't that kids' thirst for knowledge has been satiated and they now know everything they want or need to know.

The reason is that, by and large, the education system (aided and abetted by many parents and governments) doesn't celebrate, much less tap into, children's hunger to explore, inquire and discover. The system simply isn't set up to do that. Schools were designed at the turn of the nineteenth century to meet the needs of a completely different economy, which required workers who'd been equipped with a reliable, standardized package of knowledge.

Today, we need workers who are excited about learning

and know how to adapt to rapidly changing circumstances—just think of all the changes the Internet has wrought in so many different industries over the past five years—and come up with new ways to do things. We need workers who question rules of thumb and conventional wisdom, and ask, as they do at SawStop, "How do you know?"

Currently, however, most schools don't reinforce or reward divergent thinking. How can they, given their mandate? So instead of learning how to learn, many kids are learning how to be good at going to school. The straight-A student is, in virtually every educational setting, the one who has figured out what the teacher wants and how to deliver it.

My point is not that kids shouldn't be learning facts and shouldn't be memorizing, say, the letters of the alphabet or doing addition and subtraction drills. My point is that, for many kids, this is the only kind of learning that's going on at school. Their natural curiosity—the kind that keeps them excited about finding out more—gets damped down. The kids at the back of the class, the ones who challenge authority or check out altogether, may wind up scrounging for change on the street corner. But they may also be the ones who go on to start a software company or come up with a new way to treat diabetes. Whether or not schools reward "Why?" the world certainly does.

QUESTION THE UNQUESTIONABLE

Highly innovative people share five distinct "discovery skills," according to a six-year research project involving more than

500 individuals who started innovative companies—the likes of Jeff Bezos at Amazon.com and Niklas Zennström, the guy who came up with Skype—as well as 3,000 highly creative executives. The most important discovery skill, the one that can "turbocharge the others"? Asking questions. And of all the questions innovators asked, one was rated as being the most important: Why? Study participant Tata Group chairman Ratan Tata summed up the innovative mindset this way: "Question the unquestionable."

The results of the study were published in *The Innovator's DNA,* co-authored by legendary innovation guru Clayton Christensen, Jeffrey Dyer and Hal Gregersen, who point out that managers tend to ask *how* questions, like, How are we going to speed up production? Innovative entrepreneurs, on the other hand, ask Why? and Why not? They are the kids at the back of the class, all grown up: skeptical, unimpressed with conventional wisdom and pretty sure there's a better way. Michael Dell, for example, told the authors that "his idea for founding Dell Computer sprang from his asking why a computer cost five times as much as the sum of its parts. 'I would take computers apart . . . and would observe that $600 worth of parts were sold for $3,000.' In chewing over the question, he hit on his revolutionary business model."

This wasn't a particularly brilliant question. It was actually pretty basic. But asking basic questions forces people back to the heart of the matter, to re-examine and justify practices and beliefs that have become so ingrained they're almost invisible. And the payoff that comes from questioning assumptions and rules of thumb can be huge. "In business, the big prizes are found when you can ask a question that challenges the corporate orthodoxy," Andrew Cosslett, the CEO of the InterContinental Hotels Group, told the *New York Times.* "In every business I've worked in, there's been a lot of cost and value locked up in things that are deemed to be 'the way we do things around here.' So you have to talk to people and ask them, 'Why do you do that?'"

In other words, you have to ask the kinds of questions a three-year-old would ask, the kinds of questions that schools should be encouraging kids of all ages to ask. But as the educational system is currently constructed, the right answer, not the cheeky question, gets the gold star—and the faster you get that answer, the better. Pause en route to consider many different possibilities, and everyone else will whip right past you and win the race. Now, I'm all for right answers. I would not be pleased if my son came home with a gold star because he'd provided original but incorrect answers on a spelling quiz.

However, one unintended by-product of an educational system that almost exclusively rewards coming up with the right answer is children who, understandably, learn to fear giving the wrong one. Students who care about marks rush to find the answer and get the gold star, and the more gold stars they get, the more likely they are to rely on this winning formula and the more afraid they may become of making mistakes.

Yet there isn't a story of innovation or progress that doesn't involve multiple false starts and flubs. Curiosity requires the courage to risk being wrong—which, in the end, doesn't require all that much courage if you don't view being wrong as catastrophic. As British educator Ken Robinson put it in his now legendary talk as part of a Technology, Entertainment, Design—or TED—conference in June 2006, "If you're not prepared to be wrong you will never come up with anything original. And we're now running national education systems where mistakes are the worst thing you can make. And the result is that we are educating people out of their creative capacities."

Robinson, who chaired a major British government report on creativity, education and the economy, has statistical backup. A study of 1,600 children shows that divergent and nonlinear capabilities—key components of innovative thinking—drop sharply once children enter the school system. In this study, 98 percent of the three- to five-year-olds could think in divergent ways. By the time they were eight to ten years old, only 32 percent of them could think divergently. By the time they were teenagers, the percentage had dropped to 10 percent. Robinson concluded, based on this and other research, that the culprit is the educational system.

To be fair to teachers, kids really are much less compliant than they used to be, they really do grow up much faster and there really is tremendous pressure (not just from school boards but also from governments and many parents) to frogmarch kids through a standardized curriculum—all of which can translate into significant challenges in the classroom. But it's also true that teachers themselves tend to be a self-selected bunch. Many of them got into teaching because they enjoyed school and did well there; hide-and-seek pedagogy worked for them, and it's what they hope will work for their students, too. Interestingly, one recent study of prospective teachers found that the unhappier they themselves had been at school, the more likely they were to value innovative, original and unexpected thinking and to think they could help develop this in students. The prospective teachers who'd been happy at school took a dimmer view of the value of divergent thinking and were less certain that they could help kids learn to do more of it.

Many teachers believe that rote learning should start very early and that creativity is better encouraged in the higher grades. Researchers surveyed student teachers to find out when they thought it was most appropriate to focus on rote learning and memorization, as opposed to more innovative, open-ended, divergent thinking. Now, these were teachers immersed in studying educational theory, and therefore, theoretically anyway, exposed to the most cutting-edge thinking. And the majority of them thought the focus on rote learning should start in Grade 1—when children are six or seven years old, in other words, and still asking a lot of questions without worrying about looking dumb or making fools of themselves, without the self-consciousness that kicks in a little later and makes so many tweens and teens afraid to question the unquestionable.

The problem with thinking that rote learning should precede more open-ended learning is that many kids are then trained out of asking questions. By the time it's permissible, they're already forgetting how—and they've learned that answers, not questions, are the route to good grades. And to teachers' approval, which, despite older kids' posturing, really does matter to many students. While many teachers value innovative thinking in the abstract, in reality, they tell researchers that the kids they like the least are the ones they also rate as the most curious and creative. These are the kinds of children who are forever taking the class off track with their questions and observations. The kids whom teachers like the most, according to a growing body of research, are the compliant, polite, predictable ones. The kids who don't pose challenges.

THE CURIOUS THING ABOUT HIGHER LEARNING

Oh well, you may be thinking. It's too bad about elementary and high schools, but everything changes in university. That's where people are forced to stretch intellectually, so if their curiosity has been stunted, the damage can be undone.

But apparently, matters are even worse at institutions of higher learning. My source of this information? Celebrated professors at those same institutions. "The traditional way of thinking about learning at a university is that there's somebody who's a teacher who actually has some amount of knowledge, and their job is figuring out a way of communicating that knowledge to someone else," Alison Gopnik, a professor of psychology at the University of California, Berkeley, told *Maclean's*. "That's literally a medieval model, it comes from the days when there weren't a lot of printed books around, so someone read the book and explained it to everybody else. That's our model for what university education, and for that matter high school education, ought to be like. It's not a model that anybody's ever found any independent evidence for . . . I don't think there's any scientist who thinks the way we typically do university courses, where we have 300 people in a lecture hall and someone standing at the front and talking to them, has anything to do with the best methods for getting people to learn."

Gopnik went on to point out that because of the "insane" competition to get into top universities like McGill and Harvard, by the time students finally arrive, they've been trained to focus on grades rather than on taking intellectual risks or asking questions. The important thing is getting an A, not having an original thought. "We're selecting a group that has gone through so much pressure to get to university that they don't have that wide-ranging curiosity that's a really important part of having an intellectual life," she said.

If you're grooming your child for a prestigious university, it's rather unwise to cultivate curiosity. Specialization and expert knowledge is what seems to impress admissions officers. Test scores are important, too. There's a booming business in tutoring kids for standardized exams like the SATs; one woman I know told me how fortunate she felt when a top tutor in Toronto offered her seventeen-year-old daughter her last time slot: 9 p.m. on a weeknight, for $200 an hour. And then there are the extracurriculars. A kid can't simply try out the guitar then switch to the trombone and dabble in the choir; prize-winning soloists are the ones who attract the notice of top universities. Volunteering for a variety of charities is likewise not a great idea; better to start and run your own. Playing a bunch of sports? Ditto. Better to be a star at one. A narrow focus, rather than wide-ranging curiosity, is reinforced by increasingly competitive university admission guidelines (and by increasingly anxious and ambitious parents).

David Helfand, who's the president of Quest University Canada in Squamish, British Columbia, and chairs the Department of Astronomy at Columbia University in New York, has observed a big change since the 1970s. Back then, after the student protests of the 1960s, it just wasn't all that tough to get into Columbia, despite its Ivy League status. The university accepted 30 percent of the students who applied and they tended to be kids who lived in the northeastern United States. "Today Columbia accepts 6 percent, and they're from all over the world. And the quality of the students, they say, is much, much higher," Helfand told me, but he wasn't bragging. It was a lament: "Really, what's much, much, *much* higher is the willingness of students to jump over any hurdle—however high you put it—to get into this university to get the degree. We really have commoditized education. We talk about the students as customers. And they come in and they buy the education, and some of them buy the really high-priced luxury brands like the Ivy League, and some of them go to the big box stores, like the big provincial or state universities in the U.S." Whatever the brand, he continued, the prevailing attitude among students is that they are buying a degree, a ticket to the workforce, not an education.

This attitude, combined with the systemic emphasis on shut-up-and-listen-to-me lectures rather than experiential learning, won't undo any earlier damage caused by hide-and-seek pedagogy. By university, many students have learned that questions just aren't that important. What really matters are marks, and you get good ones only if you know the right answer.

THE QUESTION OF CULTURE

In Canada, parenting and schooling aren't the only forces that can dampen down curiosity. There may also be cultural barriers to asking questions. With our fortress mentality and stiff-upper-lip British heritage, we may have unintentionally created a society that in subtle ways encourages us to resist the whole notion of challenging tradition. Of course, asking a question *is* an inherent challenge to the status quo, but it doesn't have to be viewed as pushy and impolite, or as a threat. It can also be seen as an invitation: "Let's kick this idea around a little, have a dialogue, see how we can make things even better."

But I think Canadians tend to hear certain types of questions, particularly ones that don't involve the weather or a hockey game, as rude or inappropriate. I remember being slightly taken aback, when I moved to New York to work for the *Financial Post* and later CNN, by how unabashed Americans were about asking personal questions. It was not uncommon to be quizzed on everything from relationship status to career prospects within the first five minutes of meeting someone. There are nosy Canadians, to be sure, but there is a particular quality to American curiosity, a lack of apology for it, that you see much less often here, and it took a while to get used to it.

I was reminded of this recently when I spoke to a group of executives about innovation and met three American-born CEOs who now run Canadian subsidiaries of large corporations. All three are in the consumer goods space, and all had war stories about the culture shock they experienced when they first began managing Canadian employees.

Politeness, they agreed, was a real barrier to problem solving in their Canadian offices, in a way it hadn't been when they worked in the United States. "Problems are gold!" said a CEO, whom I'll call George, in a gravelly, Midwestern accent. "Making them visible should be a good thing. Instead there is a fear of hurting feelings, or seeming rude or abrupt." A leader I'll call Mike, also a Midwesterner, agreed. A reluctance to address problems head-on was also a concern at his company. He told us about a mid-level sales executive who'd run a moderately successful evergreen client file for years and was expecting to be rewarded one day with a promotion to vice-president. But Mike could see that she was someone who thrived on reliability and predictability and keeping things the same—simply not innovative enough to be a VP, and when he told her that, as gently as he could, she burst into tears. She had never asked for, nor had anyone ever volunteered, direct feedback on her prospects, much less indicated that the status quo was something she should be trying to change.

"Tom," another ex-pat CEO, nodded in recognition, adding, "Canadians tend to dance around these kinds of conversations." He coaches his son's basketball team and said he's observed a real difference here when a kid has a free throw. In the United States, he explained, supporters of the other team go crazy in that situation, making as much noise as possible to try to throw the player off and upset the throw, "but, in Canada, the whole crowd goes completely silent so the player can concentrate."

He's all for politeness and courtesy, he continued, but it does tend to make it more difficult to have direct conversations and to see questions as healthy challenges rather than personal attacks. "At my first meeting here, a strategic planning review, thirty people showed up to the boardroom," Tom said. "There should have been something like twelve. But nobody wanted to say to anyone else, 'Why are *you* coming to the meeting?'" And once there, people fell over themselves complimenting each other's ideas. "It was not useful at all. We needed them to challenge each other."

Viewing challenges as breaches of etiquette translates, all three agreed, into a "good, not great" mentality in the business world. All three saw fear of conflict and fear of making mistakes as serious stumbling blocks for their Canadian employees, and interestingly, all three noted that new Canadians seem much more likely to be willing to question conventional wisdom.

My purpose is not to say that Americans are better than Canadians—if I believed that, I'd still be living in the United States—or that we should ape their ways. My purpose is to point out that cultural factors may act as dampers on our natural curiosity in ways that we should be aware of, so we can work to counteract them—not in order to become more like Americans, but to become more successful Canadians. Curiosity doesn't just help us evolve as individuals but as a society, because it powers the kind of innovative thinking that results in new ideas, new ways of doing things and new products.

Innovation drives productivity, which to a large degree determines our standard of living at home and our competitiveness abroad. And Canadians' productivity has been tanking since the early 1970s. Today, Americans, who've always been more productive, have pulled way ahead of us. In the United States, worker output—measured as the value of the goods or services produced per hour by that worker—is $44. In Canada, it's only $35, which makes us sixteenth among the seventeen countries the Conference Board of Canada compared in a recent study. That's bad news for business—corporate profits could be 40 percent higher if we could match U.S. productivity rates. And it's bad news for government. If we closed the productivity gap, revenue to our government would increase by $66 billion a year, without raising taxes a dime. Goodbye deficits! And it's *really* bad news for you and me, because if we could match U.S. productivity rates, we'd have more in our wallets. A lot more.

So there are some very good practical reasons to encourage curiosity: it drives innovation, which in turn powers productivity. If there's something that's holding us back, culturally, we'd better try to figure out what it is and correct it. Fast.

HOW COULD WE DO THAT?

Encouraging people to ask more and better questions is not all that hard, as it turns out. While curiosity is a trait—that is, a stable characteristic more pronounced in some people than others—it is also a state, meaning that it can ramp up and down in the same way moods do. And it's a state that's pretty easy for us to encourage in young children not only by answering their questions but also by encouraging them to ask more of them.

This can be as simple as role modelling. In one experiment, student teachers were told to make a point of asking a kindergarten class a few divergent-thinking questions every day over an eight-week period. So when they were reading stories, they'd stop and ask the kids what they thought might happen next. At the art table, they'd ask, "What other sorts of things could you do with these art materials?" During snack time, they posed open-ended questions like "If you were lost in the forest, what are some things you could do?" In the control classroom, student teachers were also instructed to ask questions to extend children's learning, but they weren't specifically instructed to ask questions that called for divergent thinking.

Supervisors kept track of what types of questions teachers were asking in both classrooms, and the teachers themselves also submitted records of their questions. In the first classroom, there were more than 250 instances of divergent thinking over the two-month period, whereas in the second, there were just 25. And although both classes got roughly equal scores on a classic creativity test before the study, afterward, the kids who'd been asked divergent-thinking questions way outperformed the control group on every measure. The study's authors concluded that "very young children can realize dramatic increases [in innovative thinking] when repeatedly exposed to divergent-thinking situations" and urged that this "be encouraged at an early age."

Clearly, this is something we can also be doing as parents: modelling asking more open-ended questions, and then encouraging our kids to speculate and let their minds wander. And like Gass's parents, instead of programming our kids into structured activities round the clock, we can also encourage them to find things out for themselves.

When kids are encouraged to explore and figure things out on their own, they learn more, according to several recent studies. In one experiment at MIT, kids were given a toy that responded in a variety of ways—by making different noises and so on—depending on how it was handled. With one group of children, the experimenter held up the toy and said, "Oh look, I've never seen this before, let's see what it can do," and then handed it over to the kids. With another group, the experimenter said, "I'll show you how this toy works," and proceeded to demonstrate several of its features. The first group of children spontaneously explored and figured out every last thing the toy could do, but the second group just did exactly what the experimenter had done with the toy, over and over again.

Adults also learn more and better, and ask more and better questions, when they are allowed to explore and experience things for themselves—a hopeful indicator that it's never too late to try to kick-start our curiosity. In one study involving science educators enrolled in an eight-day professional development seminar, teachers spent forty minutes at three different inquiry stations to see what would happen depending on the level of instruction they received. Each station was equipped identically, with everything they'd need to make foam: electric mixers and other kitchen utensils, various detergents, eggs and other cooking ingredients, and toiletry articles such as shaving cream. At the structured inquiry station, teachers were given fill-in-the-blank worksheets and explicit directions for making foam; at the guided inquiry station, a supervisor posed questions to the teachers while they were experimenting with the items; and at the full inquiry station, teachers were left to their own devices and there were no instructions—and no mention, even—of producing foam.

The experimenters, who recorded what went on at each station, expressed shock at what occurred at the full inquiry station for all groups: many more spontaneous questions—twice as many as at the guided inquiry station, and about eight times as many as at the structured inquiry station—and much more probing questions. At the other stations, teachers asked each other things like "What do they want from us?" At the full inquiry station, though, teachers asked each other why different sizes of bubbles developed and how to determine the efficiency of different mixing methods—what the experimenters classified as higher-order curiosity rather than procedural questions.

Clearly, parents and teachers ought to be encouraging children to ask higher-order, curiosity-type rather than procedural questions if the goal is to turn out innovative thinkers who are engaged by learning. But ultimately, we shouldn't do that just because it might improve their career prospects and help our economy (though let's face it, those are pretty good reasons). We should do it because it will improve their lives.

While people who are capable of divergent thinking may or may not wind up in the corner office of cool, innovative companies, one outcome is almost guaranteed: they will lead richer, more interesting lives than less curious people do. Happier ones, too: curiosity is strongly linked to feelings of well-being and also believing that your life has meaning. Social scientists think this is because curiosity is all about growth and discovery. Curious people see and do things that less curious people do not, either because they don't spot the opportunity or because they are too wary to take it. Curiosity drives people to "challenge their views of self, others, and the world with an inevitable stretching of information, knowledge, and skills," leading researchers conclude. Curious types are more likely to approach, not avoid, "novel, uncertain, and complex activities." This keeps life interesting. Exciting, even.

Curious people tend to be more self-aware, too. They don't shy away from the tough questions: Who am I? Where do I want to go in life? And why? Instead of sleepwalking through life or doing what's expected, they're wondering how to get the most out of their days and the most out of themselves. This is as true of the stay-at-home parent as it is of the CEO. Curiosity doesn't determine a particular path in life; it just makes it more likely that you'll choose the one that's right for you rather than doing what you think you ought to want to do, or what will win approval from others.

So curiosity isn't just the pathway to creativity and innovation. And it isn't simply the antidote to complacency and boredom. It keeps life varied and meaningful. Curiosity is, in many ways, its own reward.

CHAPTER TWO

Forget What You Think You Know

Stephen Wetmore isn't a retailer by training—he spent decades working in telecom and health care before taking the job as CEO of one of Canada's best-known retail brands, Canadian Tire. It was, perhaps, his outsider's perspective that led to some of the most important insights Canadian Tire has had in recent years, ones that not only continue to shape its advertising and marketing strategies but have also fundamentally reshaped corporate culture.

By the time Wetmore arrived in early 2009, Canadian Tire had branched out well beyond its traditional staples—car parts, sporting goods, tools, hardware—and was making a concerted effort to appeal to women. It makes sense. If your customers are mostly male, there's plenty of growth potential if you can figure out how to get women in the door, too, particularly since research shows that women make most of the purchasing decisions in a family. So the retailer had focused on and reinvigorated its Living division—home decor, storage and organization, kitchen items—to make it more attractive to female consumers.

Nevertheless, sales were off and profits were down—way down, by almost 11 percent according to the 2009 annual report. Furthermore, Home Depot, Walmart and Lowe's had emerged as serious competitive threats. Wetmore, looking around the stores with fresh eyes and zero vested interest in the new Debbie Travis department, saw something else: male in-store traffic was also down and, on surveys, positive feedback from men was declining. A brand that had always been very masculine at heart was losing touch with its core customers. Something needed to be done, quickly, to arrest the slide.

Wetmore asked his newly founded Research and Insights team to find out why men weren't flocking to Canadian Tire. "Let's go deeper," urged Matthew Feaver, then the company's consumer insight analyst. His view was that answering the problem the typical way, with focus groups about Canadian Tire, might address the symptom but miss the cause, and therefore other, related issues could arise in the future.

The challenge, Feaver was convinced, wasn't figuring out what the retailer was doing wrong. The challenge was to figure out what men want. And this could not be done, Feaver believed, if the chain focused on finding the one right answer as quickly as possible. That would entail starting with a lot of assumptions about men—that guys like tools and cars, say, or that they don't like going shopping with women—and would make an open-minded approach almost impossible. An approach that promoted divergent thinking, Feaver argued, would wind up serving the retailer better in the long term.

“We took the initiative to say, ‘Let’s understand what’s important in men’s lives, and how they choose what they will and will not do,’” Feaver remembers. The quick-talking, energetic thirtysomething was used to rocking the boat inside companies—at least one former boss had asked him to slow *down* the idea generation so everyone else could keep up. Getting his way at Canadian Tire wasn’t easy. “It took a lot of pushing and a lot of fighting,” he says. Like most companies with a problem, Canadian Tire wanted to fix it ASAP. But Feaver’s group wasn’t afraid to challenge the conventional way of solving problems; it helped that, like Wetmore, most of them were new hires with no investment in business as usual. “We said, ‘Listen, give us a shot here, just enough rope to hang ourselves,’” Feaver recalls, “and we were able to get a budget and get the timelines that we needed.”

The mandate was to seek out a true understanding of how Canadian men—their roles, aspirations, motivations—had “evolved,” so the year-long undertaking was dubbed Project Darwin. The women who worked on it said—and still say—that it changed the way they think about and relate to men. Many men who participated said it was the first time they’d ever been able to speak openly about how it feels to be male. “Some guys in the organization said they wanted to laminate the insights and take them home, so they’re always there for their wives to see,” says one project leader.

And Project Darwin fundamentally changed the way Canadian Tire thinks about and markets itself to men.

WIPING THE SLATE CLEAN

The Project Darwin team members started with just one assumption: men were losing interest in Canadian Tire. They really had no idea why, nor did they care—for the time being. Their mission wasn't to focus on answers and conclusions but to ask questions that would delve into the psyche of Canadian males of all ages and ethnicities, as though men were a foreign species about whom nothing was known. Although they didn't formulate it in this way, the entire point of the exercise was to approach the topic in much the way small children would, by questioning what seems obvious and unquestionable: What is a man, anyway?

The team holed up in a place that itself looks a little foreign—"Retail City," just north of Toronto, where there's a full-scale mock-up of a Canadian Tire store that's used for planning and training purposes. The Project Darwin team worked in a big room that was suitably dramatic, given the sweeping nature of its mandate. It has a high ceiling, a plastic car jutting out of one wall, and stairs leading up to a catwalk fifteen feet off the floor.

The core players were Feaver; Jana Meerkamper, a bubbly woman with a boisterous laugh; and Cedric Paivin, lanky and droll. Their first step was to consult the zeitgeist: read articles and books, watch movies, screen ads—anything to help them identify broad themes and raise questions. The ones they came up with were big: What does it mean to be a man? How do men define their roles as fathers and husbands? What do they feel they're on the hook for? What are their dreams? What matters to them? What do their wives and children think of them?

You might think, given this starting point, that what followed would be a free-flowing, all-over-the-place discussion of the sort you used to have late at night in university, but the way the team set about considering these questions and looking for more was meticulously planned and carefully orchestrated well in advance. The quality of the insights it would generate would be linked, Feaver thought, to the quality of the process it used to find them.

The next step was a series of more than a dozen intimate, three-person focus groups with men of all ages, led by a professional facilitator while Project Darwin team members observed via a one-way window. The facilitator, considered one of the best in Canada, was a woman, specifically chosen for her ability to engage men. When she introduced herself, she explained that over the next two hours, she really wanted to find out about the men—what they were thinking, what their lives were like, what their passions and dreams were. When at one point she added, "This is probably not going to be the average conversation you've had with women in the past," one guy shot back, "So does that mean you're not going to finish my sentences?"

It was the perfect stepping stone to exactly the sort of exchange the team wanted to hear. Once participants knew that everything was off the record, Paivin recalls, "It was unbelievable what they started confessing. Guys don't talk about this kind of stuff, about themselves." In one focus group, a new father with a six-week-old at home kicked off the session by waxing eloquent about the joys of parenthood: "I've got this new baby, and it's so much fun!" But as the group's conversation continued, the truth came out. "I'm a musician, and I haven't really played in the last couple of months since the baby's been here. God, I miss playing my music."

On the other side of the glass, some of the Canadian Tire women were affronted. Meerkamper remembers female colleagues harrumphing, "You think it rocked *your* world to have a baby?! Well, it rocks our world, too!" Meerkamper and Paivin had a different take: this is what it sounds like when a guy is being honest about the baby blues, his malaise and discomfort with the radical shift in his life, how much he misses hanging with his buddies.

The empty nesters in the focus groups struck a more wistful and reflective note. Some had retired and were living through an unsettling role reversal. Suddenly, they were taking on more domestic responsibilities, whiling away the hours until their wives got home from work. These men talked about wanting to be more involved with their grandchildren; they really hadn't been there for their own kids, and were now thinking about what kind of legacy they were leaving.

The Project Darwin team took notes but did not yet try to categorize anything. They didn't want to leap to conclusions or go into solution mode. They were still trying to "go deeper," as Feaver put it to the CEO of Canadian Tire, to get to the why of the matter. So they gathered more information by having men blog online while, separately, their wives and kids also blogged, though neither group could see what the other was writing; this gave the team different perspectives on the same family. They also conducted what are called "shop-alongs": shadowing men in stores and scribbling notes on their habits, preferences and patterns.

And in several cities across Canada, the team hired ethnographers—experts trained to observe and interpret the systems of meaning that guide the life of a cultural group—to visit men in their homes in order to study how they lived and interacted with their families. Project Darwin team members tagged along for some of these visits, and while the families being studied knew they were researchers, they didn't know the purpose or even that it was for Canadian Tire. The idea was to be flies on the wall for six hours, just watch people do what they would normally do on a Saturday—without interacting with them or changing their routines—and write everything down. This is how Meerkamper came to find herself, one weekend, observing a couple puttering around the house and doing chores. It was a second marriage for both, and she was struck by how seamlessly the couple performed their tasks, saying little as they cleaned but moving around each other in what seemed to be a kind of choreographed dance. "She did her thing, he did his, they would sometimes meet in the hallway and kind of update each other. It was all really civil," Meerkamper said. "Wow, this is beautiful symmetry, they just seem to have a flow here," she marvelled. Trailing after them as they did their grocery shopping, it was the same story: no need for lengthy discussions, the wife just headed off to one aisle while the man hit the produce section. They seemed completely in sync.

Until the end of the session, when Meerkamper and the ethnographer split the couple up to talk about what they'd seen, and the wife immediately began listing all the mistakes her husband had made that day. "Did you *see* how he picked those vegetables? He didn't even look at them! He was too busy talking to someone. And did you *see* how he washed that bowl? I'm going to have to go back and do it properly." She had a long catalogue of his crimes, though she'd failed to point any of them out at the time they occurred. Remembering this, Meerkamper laughs, but it's a rueful laugh. "The one big learning from this whole study is that women are hard on their men. They really are," she says. When she and Paivin present the team's findings internally or externally, they usually start off by saying that every woman who's touched Project Darwin vows she is going to treat the man in her life better.

Another breakthrough finding: men are reluctant to speak ill of or even acknowledge anything negative about their spouses. Cedric Paivin remembers tagging along with a French-speaking ethnographer in suburban Montreal, who took him to the home of a lovely young family. The woman made coffee while their baby toddled about, and the man crowed, "I've got the perfect wife! I married my high school sweetheart, I've got a beautiful healthy baby boy, I'm paying down my mortgage, I've got a stable job, I love the guys I work with." Sensing this picture was a little too rosy, the ethnographer asked to speak to the husband privately in the backyard, where he promptly broke down. "I have to come clean," he wept. "I was pressured to marry my high school girlfriend, she's not as good as I told you she was and I'm not in love with her. I was pressured into having kids."

That episode raised more questions for the team: Are men different on the surface than they are inside? Do they present a face to the world that doesn't match their internal feelings? And if so, how on earth can we figure out what they're really feeling?

At this point, it might have been tempting to say "Enough already" and get to work trying to find a solution to Canadian Tire's problem, but the team was far from finished. Actually, the more the team members found out, the less they felt they knew about men and the more questions they had. "We knew it would be premature to draw conclusions from a couple of qualitative stages," explains Paivin. "They're just not representative—the samples are too small—but you get certain nuggets. The quantitative stage is really where you test the hypotheses and validate."

But they weren't there yet, and they weren't rushing things. The next stage, a workshop, brought together a team of fourteen employees from various departments across the company—merchandising, marketing, store design and so forth. The point was "insight generation" based on the findings from all the field research, and the hope was that bringing together a diverse group of people would result in even more ideas and questions. To prepare for the workshop, everyone got research packs with about forty pages of reports, statistics and articles; each pack was different, though there was some overlap of material. Employees were instructed to read all the material and, using Post-it Notes, jot down any facts that leaped out at them with a red pen and note their own questions or observations with a green pen. Everyone came to the workshop with a pile of Post-its, so many that once they'd been grouped by subject matter, they covered an entire wall of the cavernous room in Retail City. That many of the observations were contradictory bothered no one; the core team still wasn't at the stage of trying to look for right answers and pitch out those that seemed wrong.

The second pre-workshop task was for each of the fourteen participants to prepare a "consumer immersion mission," the point being to create empathy with the subjects they were studying—men, in other words—and try to see how the world looked to them. Participants could share what they'd learned in their mission any way they chose—a talk, a PowerPoint presentation, a collage, "interpretive dance, we don't care," Paivin joked—and afterward, everyone else wrote down their reactions and observations on yet more Post-its, which were added to the wall.

The diversity of the immersion missions was striking—people were interested in very different things, and thought about how to obtain information very differently. One person compared boys' and girls' toys; another researched why men cheat on their wives. Someone else reviewed television commercials, prompted by research indicating that men didn't think advertisers "got" them, and prepared a reel of representative ads for the group. "It wasn't pretty," Meerkamper says. "Men are portrayed as metrosexuals or as retrosexuals, the old-school tough guy who's interested in beer and women and sports and that's it. And the other type of guy was just a doofus, basically, who's oblivious to how badly he's performing. When you see these ads back to back, you're like, 'My goodness!'"

Which naturally led to more questions: How do men feel about being portrayed as cavemen or doofuses? Is this how they see themselves? Or is it such a turnoff that they'll abandon a brand?

Senior executives at Canadian Tire might have thought these missions were ridiculous, and Meerkamper concedes that in isolation they were. However, they provided "a new spark, a new way of looking at all that data. It just makes you think about things a little differently." Thinking divergently, not trying to converge on the one right answer, was the whole point of Project Darwin. So no one objected when Matt Feaver did an immersion with an animal trainer who works with sea lions, to find out how the males and females differ (males are showboats, willing to try a trick first, but also protective, entering the water before the females, scanning for predators). Someone else observed coed baseball leagues as well as male-only ones, and noted that the players in the single-sex teams wore full-on uniforms, used more technical language and took the game much more seriously, never calling out "Nice try" or consoling each other, as was standard practice in the coed leagues. One woman interviewed a minister to find out what guys confided, and the minister said, "Men only come to me when they're in crisis, whereas women come and say, 'I'm having a bad day.' If a guy comes to talk to me, he's in bad shape and wants to talk about one of two things: 'Is this all there is?' or 'I'm not having fun anymore.'"

So now there were new questions: What do men do for fun? How important is fun?

Variations of those questions and many of the others raised by the team's research wound up on the in-depth survey administered to 1,500 men and 500 women, which comprised the quantitative part of the research. (Spoiler alert: fun turns out to be almost like oxygen for men, many of whom want to have more of it at home, with their families—but if they can't, they will start looking for ways to escape.)

The survey was painstakingly crafted because another thing the team had learned was that wording is extremely important if you want to get men to talk about their feelings. They'd read about a survey of newly divorced people, which asked how they were coping. The women were devastated, but among the men, "Great!" was a standard response, which confirmed the stereotype that men don't have feelings and are, well, jerks. But then the survey was redone, and this time instead of asking how men were coping, it asked how they were *dealing* with divorce, and suddenly the floodgates opened, with men responding, "Oh, it's *really* hard." The takeaway: there are some words that men don't relate to—and thus there were many, many debates about word choices on the Canadian Tire quantitative survey, which was designed to unearth the thoughts men don't eagerly volunteer.

When the survey results were tabulated, analyzed and presented to the team, it was "like drinking from a firehose," Meerkamper recalls. Starting with a seemingly simple, "childish" question—"What does it mean to be a man?"—and continuing to ask it over and over, in a variety of ways, wound up yielding "so much stuff, it was almost paralyzing." All the fieldwork and the workshops and missions helped the core team interpret some findings that might otherwise have been puzzling. For instance, though more than half the wives said they nagged their husbands, only about one-fifth of men admitted that they were nagged—a gap that made sense to the team given that they'd already learned that men don't like to say negative things about their wives.

In the end, five key insights emerged that would shape—and are still shaping—how Canadian Tire markets itself to men. Though they shape advertising, they're not the core messages of ads. Rather, the Project Darwin findings now inform the way the retailer speaks to and about men.

"The Coles Notes of it," Matthew Feaver says, "is that what drives life satisfaction for men is the relationship they have with their significant other. It's uncanny, the correlation." If Canadian Tire had stopped with the question "Why are men disengaging from our brand?" no one would have figured out that a man's romantic relationship is hugely significant to his sense of well-being, or that that information could actually be hugely helpful in bringing men back to the stores. "We were messaging incorrectly in our ads," Paivin says. "We were relying on the doofus portrayal, and it became abundantly clear that's not how men want to be spoken to."

One pre-Darwin TV ad featured a couple who bicker throughout—the only thing they agree on is tires—and it tested worse among men in relationships than any ad Canadian Tire ever made. After a year of questioning, analyzing and research, the Project Darwin team knew why: men like being portrayed as people who value their families, not as doofuses who bicker with their know-it-all wives—but they wouldn't say so. The opportunity for Canadian Tire, the team realized, was to tap into ways to make men feel good about themselves, to feel like winners who connect with their families and have fun with them. So in a post-Darwin ad that debuted during the 2010 Olympic Games, there's warm music playing as a man walks into his kitchen and shows his nine-year-old son that he's bought a pair of skates. Cut to the boy lacing his own skates and venturing out on the ice, only to pause and reach back to help his dad who, we now see, is tottering on those new skates, a total beginner. It's the kind of ad that can bring a tear to your eye, and it's definitely not about selling skates. It's about men, and fatherhood, and a company that understands how complex and yet how basic love of family is in this day and age. Canadian Tire also changed its tag line, from "For days like this" to "Bring it on," to appeal to men's need for fun and their need to feel like providers and protectors.

The Project Darwin findings aren't just being used in advertising but right across the company. The broader insights about men have been shared with the merchandising/buying teams but, says Meerkamper, "the bigger impact of Project Darwin has been how it's fundamentally embedded itself in our corporate culture—one that truly respects men." And the team's method, its carefully structured yet free-ranging system of inquiry that allows for continual questioning and deep dives, has been redeployed to study other consumer groups, such as new Canadians, and to investigate other types of problems.

To me, the most powerful lesson from the project—other than how terribly we stereotype the modern male—is that true insight requires not only curiosity but patience. It takes patience to resist the temptation to start with the assumptions "everyone" knows are "true." It takes patience not to try to wrap up the question period as quickly as possible and start working on conclusions and seeking consensus. Our natural instinct, particularly when a problem is serious, is to find a fix and try to implement it right away. But the risk is that we never get to the questions that will deliver the real payoff: the big, essential insights that point to a new path forward.

WHY IS IT SO HARD TO ASK "CHILDISH" QUESTIONS?

Often the first step in innovative problem solving is to abandon preconceptions so that you can try to see the situation from a different angle. At Canadian Tire, they did this by very consciously performing the mental equivalent of pressing control-alt-delete, trying to wipe all their assumptions about men off the screen so they could focus instead on questions. By shelving conventional wisdom—men care most about their status at work; men are selfish creatures who couldn't care less what women want—they created the possibility of going somewhere completely new.

But mentally pressing control-alt-delete—those keys on a computer that force a complete reboot—at work is really difficult to do. It's human nature to want to retreat to assumptions; "knowing" is a lot more comfortable than feeling you've been cut adrift and have no clear idea where you're going. Being willing to ask big questions, especially when your livelihood is at stake and your ego is on the line, requires the courage to challenge tradition in an industry or the conventional wisdom within an organization. It requires humility to admit when you don't know something or aren't sure, and a certain degree of comfort with discomfort. You need to be willing to tolerate ambiguity and uncertainty—after all, who knows what the answers will be? You need the discipline to alter and perhaps even abandon cherished theories and plans based on new information, rather than trying to twist the new information to fit the traditional mould. Questions, of course, aren't worth a whole lot if you're not prepared to act on what you discover; the key second step is embracing the answers you uncover, even if they fly in the face of convention and reason. All of that requires an appetite for risk. You can't fear the unknown—the trick with curiosity is that you never know exactly where it will lead you.

In other words, curiosity on the job requires habits of thought that most of us don't cultivate. At school, as we've seen, but often at home also, we learned that authority figures have answers, not questions. And in the workplace, most of us would prefer to be seen as authorities—even though thinking a little more "childishly" might actually wind up making us more authoritative.

In many workplaces there are real or perceived disincentives to asking questions. A lot of people worry about revealing they don't know something; they want to look like experts, not ignoramuses. There's the fear of asking a "dumb" question, one that everyone else in the room knows the answer to, one that would make the questioner look pretty foolish.

And then there's concern about the possible consequences of challenging conventional wisdom. Most people are scrambling to secure a toehold and then hang on for dear life. Questioning isn't the focus, unless the question is how to pay the bills. We don't challenge the powers that be, no matter how we might complain about them, nor are we prone to questioning the way things are done in our workplace and how they might be improved, much less what the point of it all is. We don't want to rock the boat. And, frequently, we don't see fixing what's wrong as our problem. We've got enough on our plates already.

Often it's not apparent that employees' reluctance to ask questions is a problem for a business until there's a really big problem, like a slide in profits or a disastrous product launch. Finding solutions starts with asking questions—Why do we do it this way? How can we do it better?—that would better have been asked long before you were staring down the barrel of a gun.

Mismanagement, betting on the wrong horse, catastrophic bad luck, being overleveraged, a troubled economy, outright fraud—there are many routes to bankruptcy. But in just about every instance, there's one common thread: a failure to ask the right questions at the right time—or, alternatively, to accept the answers.

THE STATUS QUO BIAS

Part of our reluctance to embrace answers that require real change is what social scientists call the "status quo bias." As hundreds of studies in fields ranging from economics to political science to psychology have shown, people generally opt to keep things the same, or change them as little as possible, even when there are clear benefits to doing so. Whether what's on offer is a credit card with lower interest charges or a new morning routine, people tend to prefer the tried-and-true. Another cognitive bias underlies this one, and that's loss aversion. We'd rather have the security of hanging on to what we have today than the insecurity and uncertainty that come with all new opportunities.

The status quo bias can infect corporate leadership, too, sometimes fatally, even when the stakes are very high. Sometimes when a company is in trouble, its leaders are convinced they already have all the answers and refuse to rethink them. Sometimes they bury their heads in the sand, not wanting to believe that change is inevitable. Other times they have so much invested in their current business model that they simply cannot see past it and react quickly enough.

How else to explain Blockbuster, the massive incumbent player in home entertainment, missing the chance to leverage its brand and get into the low-barrier-to-entry business of streaming movies on the Internet? Failure was not a foregone conclusion for Blockbuster just because people stopped leaving their homes to rent DVDs. Think about it. Many of Blockbuster's long-time clients would have turned to an online service with a trusted brand name. But since they didn't have that option, they chose Netflix instead. Blockbuster's corporate strategy had been to sink millions into real estate, to ensure stores were handy just about everywhere in North America, but it seemed to have missed the question altogether: Why do people come to Blockbuster? It sure wasn't because we enjoyed the outing.

It was because we wanted to be entertained. Which leads directly to another question: Is there another way to get entertainment to people? Turns out there was. In 2010, the U.S. business went bankrupt, followed six months later by the Canadian division. (It's not clear that Netflix will win the fight to deliver content to homes, as the company still faces a challenge from the incumbent cable players, but at least it's still in the ring.)

Other businesses sense change is in the air and flail around, looking for a response that will suit their existing business model. They're prepared to do anything—new CEO, new marketing, you name it! Anything, that is, except question their business model. For instance, Hewlett-Packard is the world's biggest computer company, with more than $120 billion in annual sales and a global workforce of 350,000, yet is currently perceived by investors as having no idea what it's doing. Facing new competition on all fronts, the company appeared to panic, changing leaders and direction almost annually. Meg Whitman, the current CEO (at least at the time this book went to print), is the seventh in eleven years. Her immediate predecessor had floated the idea that HP might sell its personal computer business (a suggestion that Whitman, then a director on the board, endorsed). But that idea caused such consternation in the market that once she became CEO, Whitman withdrew her support. Yet another change of tack for the company. The verdict is still out on whether it will be a good one.

Even when risk is unavoidable, human beings tend to prioritize security and predictability. Roger Martin calls this the reliability bias. People want to be able to count on things, especially when there's a lot of money on the line, so they seek to mitigate risk as much as possible. Martin, the long-time dean of the Rotman School of Management at the University of Toronto, is a globally recognized leader in the field of innovative thinking; he worked closely with Procter & Gamble to develop its renowned strategy for innovation. He notes that businesses have developed a wide range of tools rooted in science and mathematical modelling to help them evaluate strategy and strike the right balance between risk and reward.

But as Martin teaches business students, the path to the future is not strewn with spreadsheets that helpfully point the way. As useful as evaluative tools can be—calculating the internal rate of return or using regression analysis (full disclosure: I don't really know what that is, either)—for deciding a project's worth, the real goal of all businesses should be to think divergently. To think of the things that *haven't* been thought of yet, or to think of better ways to do things that *have* already been thought of.

To question the status quo, that is. And it starts by going back to basics, the way Canadian Tire did.

CONTROL-ALT-DELETE IN REAL LIFE

How, in our everyday lives, can we press control-alt-delete to erase the assumptions that prevent us from finding a better way forward? Is it really possible to ignore what we "know," wipe the slate clean, turn our backs on the status quo and rethink basic questions about the world and about ourselves?

It sounds daunting, I agree. But I look at it this way: at least my problems aren't as multi-faceted and complex as the ones facing corporations. I don't need to worry about things like supply chains, distribution channels and media budgets. Mostly what I need to worry about is figuring out what my assumptions are, and then forcing myself to second-guess them.

A methodical approach to inquiry, like the one Canadian Tire used, can be replicated in everyday life. The keys are open-mindedness, focusing on questions and tolerating ambiguity as well as mistakes. One reason Project Darwin succeeded is that team members allowed opposing ideas to coexist—all those Post-it Notes!—without rushing to toss some into the garbage.

Pressing control-alt-delete requires you to embrace a new way of thinking, one that at first may feel uncomfortable if you're over the age of three. "It's a thinking mentality, as opposed to a knowing mentality," is how Matt Feaver describes it. "That's one of the biggest blocks for people: 'I don't want to have to continue to learn, I want to *know.* I want to be an expert.'"

At school, as we've seen, most of us learned to think analytically, to become people who "know" rather than people who question. And speed was of the essence. Generally, the faster you could trot out proof of knowing something, the better you did. We spent years and years in this system, so long that rapid one-right-answer thinking became a habit. And sometimes that's a good thing. It's helpful to know the multiplication tables and how to parse a sentence and the date the bomb was dropped on Hiroshima. It's helpful to know which drug to prescribe when someone is having a heart attack and which button to push if the conveyor belt is moving too quickly. Sometimes, there really *is* just one right answer.

But this way of thinking doesn't necessarily equip us well to deal with new challenges. And it doesn't necessarily help us figure out how to solve chronic or complex problems, either. Innovative solutions usually require that we mentally retrace our steps to try to figure out whether the things we've taken as givens are actually misleading us. Often, we need to reformulate questions, to "go deeper," as Matthew Feaver put it.

For more than twenty years, Claude Legrand has been training organizations and individuals to think more innovatively. Legrand, a consultant in Toronto and co-author of *Innovative Intelligence,* is a native of France who's managed to retain not only his accent but also a good amount of Gallic charm, complete with twinkling blue eyes and a courtly manner. He believes that anyone can learn to think more innovatively and claims he can demonstrate his method in as little time as fifteen minutes, but, as with improving your posture, it takes weeks or months of reminders until the habits are ingrained.

His method is founded on a simple directive: Don't conclude that the problem as it's first presented, or as you first perceive it, is indeed the *actual* problem. If you do, and you've got it wrong, the solution you produce may also be wrong. The first step to figuring out what your problem is,

Legrand says, is to deconstruct it by questioning it. To illustrate what he's talking about, in seminars he has participants write down a personal problem they want to solve. The only rule is that it has to be phrased in the form of a question and start with the words *How to.*

Recently, one woman's question was, "How to make my daughter behave better?" Legrand tackled it by first underlining each relevant word. Sometimes, he says, a seemingly innocuous word can raise a red flag. For instance, in the question "How can I come up with a solution that will satisfy both marketing teams?" the *a* is a major block. There are multiple possible solutions, not just one, but the question is now pointing you down a different (and inherently limiting) path.

With the question "How to make my daughter behave better?" Legrand decided the relevant words were *make, my, behave* and *better.* He then asked the woman to define *make.* What did it mean, exactly, given that her daughter was eight years old? What was she prepared to do to compel her? Upon consideration, the woman acknowledged that there weren't great lengths she could go to; an eight-year-old is less pliable than a four-year-old. In fact, she said as she pondered the word *make,* she remembered that in most things, her daughter learned by copying her behaviour, and therefore setting a good example might be the most powerful tool she had.

Legrand then moved on to *my,* asking gently, "Are you a single parent?" No, the woman said, already nodding in understanding, changing her daughter's behaviour wasn't going to be a solo effort—her husband would have to be involved, too.

"Right away," Legrand said later, "we were into territory that had to do with how she and her spouse interact, and different styles around child rearing." At the seminar, the woman readily conceded that she had been leaving her husband out; her approach to discipline was fairly unilateral, so there were sometimes differences in how the couple responded to the same behaviour. Confusing for their child, to say the least.

Legrand proceeded to the word *better,* asking the woman what she really meant by that. After some hemming and hawing, it became apparent that *better* meant "up to my internal, never-quite-spelled-out expectations"—an arbitrary, subjective measure that wouldn't be of much use to her eight-year-old (or to her father, who was now being promoted to partner in this behaviour-improving endeavour). At the end of this deconstruction, the woman seemed a little sheepish but happy. She had some idea that the path forward was connected to her own behaviour, her relationship with her husband and the way she communicated her expectations to her daughter.

Legrand, too, was happy, though he hadn't provided a solution. But that was never his goal. Rather, the goal was to get the woman to ask the right questions in order to find the real problems she needed to work on. And that's not easy, he says, because "we are trained to find solutions." Hunting, instead, for the right problems "is very painful for people."

CHAPTER THREE

Question Yourself

What differentiates innovative companies? The conclusion I've come to is pretty simple. Even when it feels uncomfortable or risky, they continuously question themselves—and the marketplace—and they have the courage to act on the answers, even when doing so requires radical change. Of course, whether the right questions have been asked and answered is usually clear only in hindsight. But if a company plans strategy via a disciplined process of probing questioning, the probability of making the right calls (and being able to bounce back from the wrong ones) is far greater.

The most consistently successful and innovative businesses in the world promote a culture of inquiry—not just at the top but throughout the organization. Every business seems to do this a little differently. For instance, 3M and Google give employees free time simply to dream up new possibilities and think "What if?" type thoughts. At Procter & Gamble, the power of why is understood right across the organization; the company's internal process for problem solving and product development, called Design Works, has led to the creation of such products as the Magic Eraser and the Swiffer. General Electric, which files five new patents a day on average, makes a practice of constantly analyzing and second-guessing its own corporate decisions, and is prepared to toss them when they don't pan out. For instance, bucking conventional wisdom and its own previous supply chain strategy, GE is in the process of bringing call centres back to the United States; wage inflation in India, the company believes, will make it increasingly difficult to find motivated, qualified employees.

In other businesses, a culture of inquiry comes about only as the result of a crisis. Pulling a business back from the brink of disaster always begins with what, for lack of a better word, I think of as soul-searching: How did we get here? How can we leverage our strengths to get out of this mess? And the most important question of all: Why are we in business in the first place?

Exactly these kinds of questions fuelled one of the most dramatic corporate turnarounds in modern history—IBM, which was months away from going under when Lou Gerstner took over as CEO in 1993. He made waves early on when he declared, "The last thing IBM needs right now is vision." It sounded arrogant if not insane—doesn't every company need a vision, if it's going to thrive?—but Gerstner was focused on figuring out what, exactly, IBM could already do well, not what the company aspired to do well. Gerstner was the first CEO in decades who hadn't come up through the ranks of the famously insular company; he'd been the CEO of RJR Nabisco and American Express. He didn't even have a computing background. But being an outsider gave him two significant advantages. First, he saw IBM through fresh eyes (much as Stephen Wetmore did at Canadian Tire). And second, because American Express was a customer of IBM's, as its former CEO he knew first-hand that the computer giant was failing its customers.

But he didn't know why. He just knew that customer service was not exactly a strength, and so set to work asking questions: Why do we do what we do? What do people need from us? Pretty basic questions but ones that IBM, long accustomed to being the dominant player in the mainframe computer business, wasn't in the habit of asking. And the answers led to a radical transformation. IBM capitalized on its accumulated tribal knowledge of hardware and software, and branched out to provide IT and consulting services. These kinds of services have a higher profit margin than manufacturing computers does, and they wound up saving IBM. Along the way, it must be noted, many thousands of people were fired from the company that had, famously, once promised lifetime employment. But IBM is still standing today because it was willing to question its own existence.

STRENGTHS, WEAKNESSES, OPPORTUNITIES, THREATS

Curiosity is the cornerstone of all business strategy. There are a lot of ways to come up with a strategic plan, but a good one always answers the same fundamental questions:

What do we offer that's unique? What do our customers want from us? Where are we going? How can we get there? To every question, of course, there are multiple answers, so the follow-up question, the one that helps companies weigh options, is, "Why? Why do x instead of y or z?"

It's much easier to answer these questions if a company has performed what's known in the business world as a SWOT analysis. The acronym stands for strengths, weaknesses, opportunities and threats, and the process embodies a very disciplined type of curiosity. A SWOT analysis involves asking, What are our strengths and weaknesses? What are our opportunities? What are the threats? Combined, these questions provide the answer to a fundamental question: Who are we?

This was the kind of analysis Lou Gerstner initiated when he took over at IBM: Who are we, anyway? What are we good at and what are we not so good at? He identified new opportunities—consulting—and also existing and future threats—and not just external competition but internal conservatism and complacency, too. Having this information gave him a much better idea how to leverage IBM's strengths and mitigate its weaknesses, and made it possible to reinvent the company.

If you don't know what you're good at, it's hard to know how or where to begin trying to make yourself even better, or how to take advantage of new opportunities. Self-knowledge clarifies ways to stretch and improve, by highlighting strengths that are being underutilized or could be used to achieve different ends (and revealing blind spots and weaknesses that may be derailing progress).

In order to expand, diversify and innovate, self-awareness is crucial. Without it, the risk of branching out and trying new directions is too great (as ill-prepared companies discover every year). Sometimes, a rigorous inventory of strengths and weaknesses can reveal capabilities that have been hiding in plain sight. This seems to be what happened at Montreal-based CAE, which for many years has been one of the world's largest manufacturers of flight simulators; its customer list has long included major airlines and governments. But in 2000, CAE got into a new business: providing training services for pilots, cabin crews and military aviation personnel. The catalyst for becoming an integrated end-to-end solutions provider seems to have been the recognition that its core knowledge in simulators and simulation software had a logical extension, which was that a lot of people needed to learn how to use the products it manufactured. There was a massive business opportunity, in other words, that CAE was well positioned to take advantage of given its captive customer base—and the fact that those customers wanted to focus on running an airline or an air force, not on training people how to use simulators—and the company's technical expertise in a complicated field. In 2001, only 15 percent of CAE's revenues came from training and services; today, they account for almost half of total revenues. And the company continues to branch out beyond civil and defence aviation, and now provides simulators and training in health care settings, in mines and in public safety agencies.

This transformation from equipment supplier to integrated training solutions provider couldn't have occurred unless the leaders of CAE had understood what the company was good at and what the market opportunities were. As it turned out, CAE's unique strength wasn't really in manufacturing but in understanding that people who work in technically demanding and potentially dangerous professions need to be able to practise certain skills in highly realistic settings in order to develop and maintain competence. This self-knowledge naturally suggested several questions: What else could we do to help people who already buy our products? And why should we restrict ourselves to aviation when many professions use, or could use, simulation training?

For CAE, self-awareness was the foundation for eureka. The company could not have transformed itself without recognizing "Hey, we aren't just manufacturing geeks—we're good at preparing people for tough jobs, and maybe there are more ways we could be doing that."

IF IT'S IMPORTANT IN BUSINESS, WHY NOT IN REAL LIFE?

Questioning yourself is viewed as an essential and serious order of business, not least because it sets the stage for innovation. The payoff is quantifiable, and the costs of failing to do it are potentially catastrophic.

So it's strange that in everyday life, asking "Who am I?" comes off as hippyish or smacks of navel-gazing. Many people seem to have the idea that the question should already have been answered, definitively, by the time we entered the workforce. But we began the long march to the workforce in adolescence, when we started choosing what, where and how hard to study. How many of us started down a certain path propelled not by a deep understanding of ourselves but by a vague sense of what the world valued and what others expected of us? The majority, I'd guess. It's rare to know yourself so well as a teenager or young adult that you beetle along purposefully, pursuing your heart's desire with the certain knowledge that you're perfectly suited for the path you've chosen.

And even if you did, it's quite possible you were wrong. I certainly was. From the time I was nine years old, I wanted to be an architect. Even as a little kid, one of my favourite pastimes was to draw buildings in plan view, seen from above. I wasn't a great artist by any means, but I was a doodler of the first order, prone to covering pages with shapes and lines while daydreaming. And one thing I particularly loved to do was to draw a structure—a barn, for instance, complete with horse stalls and hay loft—then figure out how to transform it into something else, like a house. The stall might become a child's room; the hay loft, a living room. Finally, I'd add people, and try to figure out how they'd use the space, which areas they'd like best and what features of the structure might be problematic for them.

When adults asked what I wanted to be when I grew up, my answer was always the same: "An architect." Invariably, the response was encouraging. As I got older, I was given books on the subject, and envisioned myself building a great bridge, the kind that tests the boundaries of physics, or perhaps an airport, something a little grander than we had in Winnipeg. My twin sister, Adrian, and I were the youngest of seven children, and by the time we were born our parents were deep into political life and fiercely committed to public service. There was always an unspoken understanding that we were to do something purposeful with our lives, to choose a direction and pursue it rather than let life happen to us. My mother and father, therefore, encouraged me to view architecture not as a daydream but as a concrete goal, and my role in a family of lawyers was to be The Architect, less because I'd demonstrated talent in this area than because I'd declared an intention. My older brother Gregory was incredibly creative in a highly technical way—he sat in his room drafting designs for fun, but nobody suggested he should study architecture, and he never advanced the idea either. And from an early age Adrian was clearly very skilled, artistically, but she wasn't encouraged to become an artist—anyway, she leaned toward law, which suited the family narrative perfectly.

So when I finished high school, I knew exactly where I was heading: the Faculty of Architecture at the University of Manitoba. I'd been planning on it for years by that point. And as you might imagine, given that fact, I did well in all my subjects. Except one. In design, the core of the program, I routinely achieved second-rate results. Not only did my classmates do better work, but they enjoyed doing it a whole lot more than I did. They spent hours in the studio, where pulling all-nighters became a badge of honour, while I escaped as soon as I could to work at home in isolation, where no one would observe me struggling. And while there, I escaped from the work as much as I could; if a deadline loomed, it suddenly seemed pressing to bake something, or clean the bathroom or paint the kitchen cabinets.

Our professors could be quite harsh, so I braced for blistering criticism when our designs were being critiqued publicly. But I was never singled out. Maybe they figured there was nothing that more talented students could learn from a dissection of my unremarkable models and drawings, or perhaps they knew that no criticism would be constructive for me. I'd already improved where I could—neatness of drafting, watercolour skills—but the core of design, the marriage of function and aesthetics, simply eluded me.

I'd always done pretty well academically without having to kill myself, so pulling in middling marks even when I *did* kill myself was hard to accept. I started to feel sick all the time—migraines, heartburn, cold after cold. And yet I avoided the obvious conclusion that architecture wasn't for me. This information threatened all my ideas about myself. Who was I, after all, if not The Architect in my family?

So after getting my bachelor's degree I was at a bit of a loss. Clearly, I was never going to be a brilliant architect. Even respectable mediocrity would be a stretch. But what else could I do? What was I even good at? I had no clue. I decided to take a year off and go to Toronto, where my mother and stepfather lived. My plan was to work for a while, then go back to school for a master's degree—in architecture, of course. Like many people who get stuck on an idea, I just couldn't let this one go. Maybe more courses would magically unleash the hidden designer inside me.

When I got to Toronto and began job-hunting, it was 1991, mid-recession, but my choice to drop Latin in Grade 9 and take up typing paid off. I was fast enough to land a secretarial-type job at the *Globe and Mail*'s electronic database unit, then called InfoGlobe. At that point, journalism wasn't even on my radar—a plus, since my job had nothing to do with reporting. I was essentially office girl Friday. My title wasn't "receptionist," but when the phone rang, I answered it. As I recall, my other duties involved printing documents and a great deal of stapling.

Although my job didn't excite me, I did love spending time in the Globe newsroom. In the early 1990s, it looked like the set of *All the President's Men*: slightly grimy, thrumming with activity that picked up urgency and volume as the day wore on and deadlines grew closer. Every evening in a huge five-storey room at the back of the building, the monster printing press would grumble to life, work its way up to a roar and spit out fresh, damp, inky newspapers. The outward-looking focus of the newsroom, the relentless drive to find out what was going on in the rest of the world, was exhilarating.

After a few months, I built up the nerve to pitch a story idea. I was thrilled, and a little shocked, when an editor gave me the assignment. I'll never forget staying up late, working on that first article. It felt so different from the all-nighters I'd pulled in school, working on architectural models that never came out quite right. There, the problem was always the same. I simply didn't have a clear picture in my head to articulate. Sometimes I would get lucky and come up with an interesting shape, but it wasn't based on any core idea or organizing principle. It was completely haphazard. Most of the time I was spinning my wheels, chasing a concept just outside my grasp.

To work hard at something challenging that *is* within your grasp is a very different feeling. You no longer feel like you're slogging through the bush without mosquito repellent, a compass or any clue where you're going. If your mind is engaged, curiosity pushes you on and keeps you going—curiosity, after all, simply entails being open to and interested in something. And if your interest level is high enough, it doesn't even feel like you're working exactly. That's how it was for me when I wrote. I liked trying to figure out why something had happened and what the consequences might be, and I liked figuring out how the pieces of a story fit together. I started freelancing for the paper, writing articles after-hours, and then, a stroke of luck. The Globe was experimenting with a classroom edition designed for schools, and I got the chance to work on it. The idea of going back to university to study architecture—which I'd still been telling myself, albeit increasingly half-heartedly, was my plan—went right out the window.

Once I'd made the leap to journalism, I hardly reflected on it. If anyone asked why I'd given up architecture, I had a pat response: "I figured out I was never going to be very good at it." It was only very recently, describing my path to a new friend over lunch, that I actually remembered, in a visceral way, being nine years old and drawing those plan views of buildings, and then painstakingly adding tiny people, complete with names, ages and quirks, and trying to figure out how they'd live in the space. At some point when I was a child, someone must have told me that what I was doing was architecture, and I came to believe passionately that architecture was what I loved. But over lunch, when my friend asked me to describe these buildings I used to draw, it finally dawned on me. I had never ever actually drawn the outside of a building. It didn't matter to me what the exterior looked like. That wasn't the point. The point was the people inside, how they interacted. Their story, in other words. In reality, what I was doing was never architecture at all. It was a whole lot closer to journalism.

In retrospect, it seems strange that I didn't recognize this earlier. Even while I was floundering at university, I didn't realize that the subject I was studying didn't bear much relation to what actually interested me. Stranger still, for someone who now makes a living asking questions, I never thought to ask myself some pretty basic ones: If I love architecture, why do I feel so miserable? If this is the right career for me, why am I not very good at it? Why do I want to spend my life doing this, anyway?

I didn't ask because I'd already decided what I wanted to be when I grew up, and it never occurred to me to second-guess myself in light of my actual experiences. Everyone else seemed to think I had a good answer to the what-do-you-want-to-be-when-you-grow-up question—"An architect" never failed to elicit approval. I'd become so accustomed to thinking this was my destiny that I ignored the mounting evidence that I'd fundamentally misunderstood what architecture was all about and was neither particularly good at it nor even all that interested in it. I knew the right answer, and I was sticking with it.

This is a trap many of us fall into, and not just when we're twenty-two years old and trying to figure out what we want to be when we grow up. On the job and at home, many of us hit on an answer that sounds "right," or that others approve of, then just stop questioning. As Claude Legrand points out in his innovative intelligence seminars, this is why many of us waste a lot of time banging our heads on the proverbial wall. "Instead of going back and looking at the question, people tinker with the solution, trying to make it fit," he explains. Even when we're unhappy, many of us don't step back and perform the SWOT analysis that would help us figure out, before we waste even more of our lives, who we really are and what we want and are capable of achieving. The relationship that's never going to work, the job that just isn't satisfying no matter what you do, the degree in architecture that is not in fact the path to happiness—to get out of these quandaries (or better yet, prevent getting into them) we need to be clear-eyed and objective about ourselves and our circumstances.

The consequences of failing to do that are the same as those facing businesses—even more dire, perhaps, because what's being squandered isn't just the potential for profits. It's the potential for happiness. We miss opportunities to innovate and to make positive changes in our lives when we aren't willing to question ourselves.

ME, INC.

Questioning our own decisions and beliefs about ourselves is an unsettling prospect. What if we discover we were wrong and now have to make a lot of changes? Some people are scared to look too closely at themselves because they're convinced that they can't change, that they simply lack the capacity to innovate in their own lives. One of the biggest impediments to innovative thinking is simply that many of us don't believe we're capable of it.

This is where someone like Rolf Smith comes in. He spent twenty-four years in the Air Force, mostly in strategic planning in intelligence, before heading to the private sector, where his mission is to help people come up with big ideas and better solutions. Now in his early seventies and endlessly energetic, "Colonel Innovation" is well known among Fortune 500 companies for his retreats and group "thinking expeditions." They really *are* expeditions, with an adventure/roughing-it angle. Rock climbing, for instance, is usually on the agenda, because Smith likes to push people out of their comfort zones and force them to see themselves differently. "As people progress in their climbing ability, they start thinking differently about what they're capable of doing," he says, "and when they get done, they realize that they've done what they considered impossible. If they were able to do a 600-foot climb, well, they can walk through walls when they get back home."

Early on, he discovered that getting people to innovatively approach professional problems, particularly chronic problems, required just such a mind shift. "They need to think of themselves as innovators, mentally incorporate themselves as innovators," he explains. So his expeditions now involve a component he calls Me, Inc. Each participant starts by compiling an inventory of his or her skills and abilities, and then sits down with a "board of directors"—three other people on the expedition, usually colleagues—to talk about something that person really loves to do. The board members provide feedback, highlighting the skills and strengths that must be required to do that thing. The point of the exercise is to reveal that this thing you love to do, which you never thought of as being particularly innovative, does involve innovation skills.

Next, each participant has to come up with 101 goals and wishes. There's a boot camp element to this: "You start it in the morning, and you don't go to sleep until you have 101," Smith says briskly. He's big on shaking up people's routines—keeping them up late, feeding them irregularly, serving beer at breakfast—to get people used to the idea and experience of change.

Coming up with 101 goals and wishes is, he says, really tough for most people. Oh, the first twenty-five to thirty are pretty easy, but "when you push beyond that, you start thinking about things that you don't really think are possible, or that you think you'd be unlikely to be able to do. That'll get you to sixty or eighty. And at that point, you start putting some *really* crazy stuff in." Almost invariably, however, when participants review and analyze the list with the board of directors of Me, Inc., they realize that some of those ideas aren't so crazy after all. When Smith tells them to narrow the list down to their top five goals and wishes, almost invariably, some of the "crazy" late-in-the-day additions make the final cut, and they're always not just about change but about innovations for that participant personally.

"Over a period of five or six days you're constantly interacting with the other people in the class about your Me, Inc., and they're pointing things out to you about your own capabilities. By the time you come up with your top five, you're not just picking the easy stuff. Because you see yourself as an innovator," says Smith. "And out of that comes your overall mission, which could be 'To create a better family life' or 'A different family life.'" Better and different are, to Smith, key indicators of personality type and innovation style. "Better," like "improve," indicates that you're adaptive. "Different" pegs you as more of an originator. Finally, with input from your board of directors, you figure out what strategies you'll need to implement your mission and come up with an action plan.

Smith's parting instruction to clients is to cross out their title on their business cards and write in "Innovator" instead, so they have to explain what that means when they hand out the cards. It's an effective way, he says, of forcing yourself to realize that the capacity for innovation is in your DNA, just like the curiosity that drives it. But that realization is just the beginning. Thereafter, you have to "practise, practise, practise," Smith says. On his thinking expeditions, he has people carry little slips of blue paper "to capture ideas." They serve the same function that Post-its did for the Project Darwin team at Canadian Tire, communicating an expectation—I'm going to have some ideas today—and ensuring that nothing gets lost in the shuffle. It's one way Smith gauges buy-in at the end of a session: How many people have made carrying and using blue slips a habit? In his opinion, electronic methods aren't nearly as effective because they're "not as reinforcing, and they're harder to play with. It's not as easy to connect two separate documents on a computer as it is to put two pieces of paper together in a different way."

Smith is all about on-the-ground applications, not high-flown theory, which is one reason he put himself through Me, Inc. years ago, so he could see what it was really like. One of his top five goals was to become fluent in Spanish, so he went to Guatemala to take an intensive Spanish course and can now get along in the language pretty well. Once, he even gave a sixty-minute talk in Argentina to 4,000 Latin American educators—in Spanish. He really wanted to get it right, so he practised with a neighbour who teaches Spanish in junior high to translate his slides, and he worked hard on pronunciation beforehand.

"During the conference I listened to everybody else, tuned my ear to the language, made some changes to my slides, and by the time it was my turn to speak I was ready to rock and roll," Smith remembers. "But my computer was incompatible with the Argentinean system, so I was standing up on this huge stage, apologizing to them, and I did that in Spanish. Then I looked behind me, and I saw that an image of me was being projected on two big screens. So I grabbed my computer, set it down on the floor, and had them focus the camera on the computer instead of on me, so it projected my slides up to the screen." That, Rolf Smith says, is what innovative problem solving is all about—"Going from 'My God, I can't believe this is happening' to 'This is impossible, my computer can't communicate' to 'Hey, we did it!'"

When you accomplish something like that, Smith continues, no matter how pedestrian the problem you solved, it's important to stop and consciously note that you've just been innovating and creating something new. It's especially important to do this in your personal life, he points out, because if you feel you've been successful at innovating there, you're more likely to feel you can do it at work, too. One of his own home innovation projects? In an effort to reduce the killer monthly air-conditioning bill when he and his family lived in Houston, he offered his daughters a cash incentive to decrease their reliance on the air conditioner (a.k.a. turn the darned thing off more), giving them half of every dollar the family saved on the electricity bill. They figured out, he says with a laugh, how to make a nice bundle that way. It was a win-win.

In *The 7 Levels of Change*—writing a book was another item on his top five Me, Inc. list—Smith explains that there are numerous levels, or degrees, of innovation. The first is doing the right things to get where you need to go; the second, doing things right; and the third, doing things even better. These first steps are all about competence and setting the stage for innovation.

Many companies that seek out his services are quite content to stop right there. Exxon, for instance, insisted it just wanted to find more efficient and better ways to get natural gas out of Alaska. It wasn't in the market for level four—doing away with things—or level five—doing things others are doing. No, level three was plenty, thank you very much. "But," says Smith, "the breakthrough idea they came up with on their thinking expedition was level 4: stop building roads in Alaska, because it's one of the big-cost items, about $50,000 a mile I think. If you can eliminate roads into a production site, that's a really big deal." Exxon's innovation was to forget the roads and use helicopters and all-terrain vehicles instead. That turned out to be a game-changer for the company.

Level six, doing things no one else is doing, and seven, doing things that can't be done, are, as you might imagine, harder. But the first step to getting there, or anywhere, Smith believes, is Me, Inc. And the main question it answers is "Who am I?"

DISCIPLINED NAVEL-GAZING IS A GOOD IDEA

Asking "Who am I?" is one of the simplest and yet most important questions that lead to success, whether you go about answering it with a thinking expedition or a SWOT analysis. Embracing the answer—even if that requires dispensing with cherished illusions—clearly helps with goal setting. It's hard to figure out where you want to end up if you don't know what your strengths and weaknesses are.

And sooner or later, life will test your resolve about the choices you've made. You may find yourself living in a place you dislike and realize you never really thought through why you should settle there. Or you may achieve a major career goal or financial goal and ask, as some of the men in Canadian Tire's Project Darwin did, is this all there is? If you find yourself shrinking away from problems, or viewing them as insurmountable obstacles rather than somewhat intriguing challenges, the problem may well be that you still don't really know where you want to go. Or why.

You know you're headed in the right direction if you feel curious and engaged even when facing difficult problems and challenges. Think of Steve Gass at SawStop—when he encountered difficulties, he never thought, "Gee, I really don't know why I still bother. This is just way too unpleasant." Instead, he dug deeper, and the reason wasn't just personal fortitude. The reason was that he was very clear on what was important to him and therefore also clear where he wanted to end up, which helped him see the potential in every problem the company faced and remain interested in finding solutions.

Successful companies, and businesses that achieve major self-transformations, are very clear on where they want to wind up. They put in the time at the front end, weighing pros and cons and fully investigating the options in light of their weaknesses and strengths. In everyday life, most of us would wind up feeling more fulfilled and happy if we, too, put in the time at the front end, and resisted the urge to rush off in a particular direction without asking ourselves, repeatedly, why we wanted to go there in the first place. Sometimes, everything works out. But often enough, it doesn't. The man who gets married because he thinks he ought to, the woman who becomes a doctor because that's what her father wanted her to be—when it's too late to achieve their hearts' desires, they can wind up asking, How did I get here?

CHAPTER FOUR

Figure Out What No One Else Is Doing

Now seventy-two, Gordon Eberts, tall and lean, walks a little slower than he once did, but he still has a hard time slowing down his mind and speech so the rest of us can keep up. When his mind is really racing, he's learned to switch to speaking in French—he grew up in Montreal—because then he has to *think* in French, and that slows down his brain just enough that it's easier for regular people to follow along. He was pretty young when he realized, to put it simply, that he could see things others couldn't—and wasn't shy about questioning conventional wisdom, even in kindergarten, where he got top marks in everything except "conduct," in which he earned an F.

When I knocked on the door of his airy midtown Toronto condominium one bright weekend morning, he opened it, crowing, "I'm definitely thinking with both sides of my brain today!" Eberts often speaks of his "neural network" and is convinced that he thinks most clearly when his right/creative side and his left/mathematical side are both "on." In the heyday of Gordon Capital, a boutique investment bank he co-founded, they certainly were. Critics might say the firm strode rather close to the line of legality, but a more accurate description would be that Eberts came up with so many new ideas and products that it was all regulators could do to keep up. Many times, regulators didn't even realize they were in need of a rule until Gordon Capital made that clear by going where nobody else had. If there was an unspoken motto at the brash young firm, it had to be "Why Not?" (though judging from the colourful nature of the war stories, there would have been an expletive in there, too; the firm was legendary for its wildness off the trading floor, but Eberts swears he took no part in any of that).

While Jim Connacher was the flamboyant face of the firm, those who worked on the street knew Gordon Eberts as the great mind behind the scenes. Eberts—handsome, with a shock of thick, steel-grey hair, and always dapper in bespoke suits—didn't particularly care about name recognition. What he liked to do—and still likes to do to this day for a select group of friends—was come up with new ideas and products. Innovation comes rather easily when you have a bloodhound's nose for hitherto undetected loopholes.

Today, stocks and bonds are regularly sold in what are called "bought deals." The issuer—a company that wants to raise capital—sells a chunk of its stock to an investment bank in exchange for cash up front; the bank then tries to sell its shares to clients for a higher price than it paid for them. If the clients aren't interested, well, the bank is stuck with any shares it can't unload. This has been standard practice in North America for decades now, and it was Gordon Capital that invented the bought deal, in the early 1980s.

Before Gordon Capital turned the system upside down, all the risk of selling stocks (or bonds) lay with the issuer, the company looking to raise money. The big Canadian brokerages banded together to form a syndicate and divvy up the available stock, which, for a fee, they agreed to try to sell to their clients. But if they couldn't, they simply pocketed the fee and returned the stock to the issuing company. In other words, the company that had issued shares bore all the risk. The syndicate brokers were a clubby bunch and not looking for new members when Gordon Capital came on the scene in 1969. Eberts and company couldn't get a foot in the door. Oh, occasionally, white-shoe firms like Dominion Securities would let Gordon Capital have a taste of a transaction, but just a tiny sliver.

So Gordon Capital decided to turn its attention to a group no one else seemed to care about: customers. In finance, there's the sell side—companies (and their banks) that issue stocks and bonds—and the buy side, meaning mutual funds, big investors, and even retail investors like you and me, who wind up buying them. In the 1970s, all the money was on the sell side, so that's where the big brokerages focused their attention. But Gordon Capital, frozen out of that club, turned its attention to the buy side, and became intimately familiar with the balance sheets and portfolios of some of the biggest institutional and pension investors in Canada. That, explains Eberts, allowed the upstart brokerage to take risks—calculated ones—that the big, old-school brokerages could not. They just didn't know enough about the buyers. Their attitude seemed to be, Why bother? Business as usual works quite well for us, thank you very much.

Eberts recalls a deal he and Connacher did with Bank of Nova Scotia in the early 1980s when the bought deal was in its infancy. The bank had $100 million of convertible bonds, a liability on its balance sheet, which it wanted to convert to stock. To do that, the bank either had to persuade all the bondholders to agree, or had to buy them out of their positions at great cost.

Eberts and Connacher rode the elevator—operated by a woman named Rose, who wore a dainty lace cap—up to the offices of Ced Ritchie, CEO of the bank. As ever, Ritchie was smoking, ashes spilling down the front of his jacket, and drinks were offered round. Eberts made the pitch: if Ritchie gave Gordon Capital the business, Gordon Capital would assume *all* the risk, effectively buying the bonds itself. In order to buy out any owners who were reluctant to convert their bonds to stocks, Gordon Capital would offer a small premium, an extra 50 cents per bond. Eberts had done the math on the transaction, and worst-case scenario, assuming that every bondholder demanded the 50-cent premium—and why wouldn't they, once they knew it was on offer?—Gordon Capital would have downside risk of about 5 million BNS (Bank of Nova Scotia) shares. However, Eberts was confident the firm could sell the stock and wouldn't be stuck with it long term.

Meanwhile, Gordon Capital's fee was very low compared to what the syndicates charged. So low that, swearing and belting back his drink, Ritchie got on the phone to see if any old-school brokerage would match the fee. Tony

Fell at Dominion Securities didn't hesitate—he wouldn't do that deal at any price because he couldn't see a way to justify the risk. Next Ritchie called brokerage firm Wood Gundy; it said it'd do the deal, for a much higher fee. Gordon Capital won the day, but Connacher wasn't celebrating. As they walked out of Ritchie's office, knowing they had to sell millions of dollars of BNS stock, he turned to Eberts, ashen-faced, and said, "I hope to Christ you understand this. Because nobody else does."

But Bay Street did understand one thing perfectly well: the bought deal, where the risk was borne by the banks and not by the issuing company, had arrived. After Gordon Capital introduced the concept, it became de rigueur throughout North America. It was one of those innovations that caught on virtually overnight.

Nevertheless, overnight success was preceded by years of practice and experimentation. The genesis of the bought deal came almost a decade earlier. To raise money, issuing companies would occasionally do a private placement, selling stock directly to investors and bypassing the market; it's a bit like selling your home privately, or selling a couch on eBay, except that in a private placement, the company might sell to more than one buyer. To win that business, Gordon Capital began offering to assume all the risk of the transaction. The firm would pay the issuing company cash and take the stock onto its own books, with the hope of interesting buy-side investor clients (those pension funds and other big investors) and thereby generating a fee. You could only take such a risk if you were pretty sure it would pan out, and the only way to be pretty sure was to have an intimate understanding of clients' current investments and needs. The reputable old brokers, steeped in history and complacency, just hadn't taken the time to get to know the investors on the buy side. Gordon Capital exploited that knowledge gap and, year by year, came to understand buy-side investors' needs better and better.

That history made the investment bank comfortable with the concept of the bought deal and taught its sales team how to persuade clients to buy new issues of stock. Of course, there was some consternation along the way, like when Eberts got back from Ritchie's office at the Bank of Nova Scotia and had to call his head trader, a man named Don Bainbridge with whom he had an already somewhat fractious relationship, to tell him they'd soon own a whack of BNS stock. "You fucking did *what*?" Bainy roared. "Is that upstairs or downstairs?" He was asking who would bear the brunt of any losses: the traders downstairs on the floor, whose bonuses might be affected, or the executives upstairs. "Upstairs, Bainy, upstairs," Eberts assured him. It was in the end a moot point. Gordon Capital didn't own the stock very long before selling it at a profit.

They may have come across as cowboys—the stories about booze and women are not exactly politically correct—but they were persuasive ones. Focusing on buy-side customers helped the firm cultivate the kind of investors who could help bankroll its bought deals, which required access to a lot of capital. Between 1980 and 1995, Gordon Capital was the highest volume trader on the Toronto Stock Exchange every year, some days hitting 50 percent of total trading volume.

In order to compete, rival brokers needed deep-pocketed partners. Eventually, every one of the old club of brokerages was bought by a large Canadian bank, and when that happened, Gordon Capital was up against behemoths and found that now *it* couldn't compete. In 1998, after a failed merger attempt with Wood Gundy and a series of setbacks, including censure of Connacher by the Ontario Securities Commission for allegedly misstating financials, Gordon Capital, a shadow of its former self, was bought by HSBC. But it left a significant legacy: a variety of financial products that are now industry standards, all driven by Gordon Capital's knowledge of exactly what mattered to its customers.

Gordon Eberts was able to innovate by figuring out what no one else was doing. When a door was slammed in his face, he didn't keep hammering on it, begging for entrance, nor did he bemoan his bad luck and look for another kind of work. He looked for an open window and climbed right in.

FIRST, SPOT THE PROBLEM

Most innovation starts with the identification of a problem. The environmentally friendly paper coffee cup is too hot to pick up—ergo, the cardboard sleeve.

But spotting a problem isn't as easy as it sounds. A lot of us unquestioningly accept the status quo, problematic though it may be. For instance, every day at 4:40 p.m. in our office at the CBC, the cleaner—a very nice person who does a good job—comes to empty our garbage cans. Which is great, except that we go on air at 5 p.m. and those last twenty minutes are pretty crucial for me. I'm usually frantically writing and researching, and I really need to focus in order to do that, but where my garbage can is situated means I have to get up from my desk in order for it to be emptied. It's a daily inconvenience that affects my ability to do my job, and I'm sure it's inconvenient for the cleaner also. It would be easier all the way around if garbage cans were dealt with twenty minutes later. But for the longest time, I didn't even consciously articulate this as a problem. It was just one of those minor irritations that occurs on a daily basis. Only recently did I think, wait a minute—I can do something about this, like finding out who schedules maintenance workers and speak to that person (who, almost certainly, is unaware of this problem).

Many people are blind to certain sorts of problems, or at least blind to the possibility of addressing them. We see some problems as just part of the fabric of everyday life, inevitable and immutable, and miss spotting the potential to do something about them. We look at a table saw and say, "Well, of course it's dangerous. It's a *saw*!" An innovator looks at the saw and sees that no one else is making it safe—not only a problem but an opportunity.

We can get used to problems and become complacent about them, even when we're dissatisfied. Until there's a crisis, or until life events force a decision point, many people live with problems that have a real impact on their day-to-day lives without thinking deeply about them, and therefore miss asking the questions that could help them get to solutions.

As Claude Legrand points out, sometimes the problem is that we've misidentified a problem. For instance, in one of his seminars recently, a man raised an issue he said was plaguing him. Every night, he struggled mightily to get his three-year-old to go to bed. His goal was to have her asleep by eight o'clock, but that rarely happened. His days with his daughter inevitably ended with frustration and irritation and the feeling that he was failing as a parent. So, he asked, what did he need to do differently to be a better dad?

Legrand responded with a question of his own: "Why doesn't your daughter go to sleep?" After a few more "Why" questions, it became clear that the problem was not the guy's lack of parenting skills, nor was it that he had a behaviourally troubled toddler. The real issue was that the man's wife regularly got home from work at 7:45 p.m., eager to spend time with their daughter, who was then unable to rev down after an exciting reunion with her mom. The problem was the family's schedule, which was why anything the father did to try to solve the sleep issue—rewarding his daughter for staying in bed, moving bedtime rituals up an hour to build in more relaxation time before eight o'clock—was never going to work. Either his wife needed to come home earlier or both parents needed to agree to a later bedtime. The important question he needed to ask himself wasn't "Why can't I get this kid to sleep?" but rather "Why does my wife work late?" And then, "Why do I put such high priority on getting our daughter to sleep by eight?"

Only when the correct problem has been identified is an innovative solution possible. Sometimes it's as simple as figuring out what no one else is doing, as Gordon Eberts did. Other times, innovators get even more ambitious: they don't want to just address a gap in the market, they want to do so with such flair that they take it over. That's what Chip Wilson did.

SET THE GOLD STANDARD

When Lululemon Athletica opened its doors in a small retail space in Kitsilano, British Columbia, in 1998, the odds that it would last more than a year or two weren't high. After all, competing in women's apparel puts you up against some of the biggest manufacturers on the planet, the ones with seemingly endless access to cheap labour and also cheap material, thanks to high-volume fabric purchases. Competing in the athletic apparel space is even harder—you're up against all of that, plus first-rate marketing teams. You're up against *Nike,* for God's sake. Good luck.

But it wasn't luck that turned Lululemon into a powerhouse and put its founder, Chip Wilson, on *Forbes*'s billionaire list. It happened, you might say, by design.

Lululemon wasn't just another small retailer hoping to take a tiny sliver of Nike's market. Wilson wasn't thinking about Nike at all. His previous company, Westbeach Sports, which he'd founded in 1979 at the age of twenty-three, grew out of his love of all board sports and his conviction that the rest of his generation was going to "rad out" right along with him. He spotted the snowboarding trend before it was obvious to the mainstream, and his company grew right along with the groundswell of interest in the new sport.

He sold that business in 1997 to Intrawest, but it wasn't just a good offer that motivated him. Wilson may come across as a classic surfer dude, with a hippyish, slightly blissed-out demeanour, but he's a shrewd forecaster of the zeitgeist. Just as he was beginning to gravitate to yoga for its combination of athleticism and spiritualism—and the relief it provided from the physical wear and tear of board sports—he thought his contemporaries might do the same. And he noted two interesting things about yoga: many women were already passionate about it, but most of them wore cotton clothing to classes and wound up perspiring right through it. Instead of looking very Zen afterward, they looked bedraggled and uncomfortable.

No one was making yoga clothes that suited the aesthetic or the demands of the exercise. Wilson decided he could do something about that. If you've ever worn Lululemon pants, you know that he succeeded. Lulus are different from other athletic gear. They actually improve the appearance of the wearer, which is why many women, including actresses like Megan Fox, began wearing them round the clock. Lululemon isn't just selling yoga pants, it's selling a better-looking version of you (just like another recent entrant to the *Forbes*'s billionaire list, Sara Blakely, the woman behind Spanx undergarments).

What's their secret? I asked Alexander Manu, a Romanian-born expert on innovation and design who teaches at the Rotman School of Management at the University of Toronto and at OCAD University. He studied a pair of pants and, within minutes, began reeling off a long list of the ways in which Lulus are just *better*. They're triple-stitched for maximum comfort, using a complex zigzag stitch rather than the standard straight one. Also, there is just one seam, not two, and it's on the inside of the leg, which is more comfortable and helps the material hug the leg better than if it were on the outside. The waistband on the pants is wider than normal, meaning it's more expensive, as elastic is costly, but also makes the pants more attractive—and slimming, as women everywhere have noticed. The construction of the pants is also more complex than normal, using up to 50 percent more panels (pieces of material) per pair, which creates the firming/holding magic that makes the rest of us feel ever so slightly more like Megan Fox when we wear them. Then there's the fabric itself, a state-of-the-art blend that includes luon, as opposed to the conventional spandex that competing brands use. All in all, Manu explained, Lululemon pants are superior in terms of fabrication and execution.

Clearly, Chip Wilson and his designers figured out how to do something no one else was doing: make yoga clothing both functional and fashionable. But it's more than that. They've created a corporate culture that's quality-obsessed. Take a glance at the job openings listed on the company's website. Raw material manager, garment developer—the company is constantly innovating to improve its designs and products, which allows Lululemon to keep resetting the bar on quality yoga pants. The pants are so well made that they last a long time, but consumers go buy new ones anyway—out of pure desire, not need (my wallet can testify to this, I'm afraid). That's rare in the utilitarian sports gear market and creates a real problem for manufacturers at the low end of the market, who are pumping out $20 yoga pants. The

Lululemon customer might try them on, looking for a bargain, then turn up her nose: they're just not as comfortable. Plus, she just doesn't look as good in them.

But why doesn't Nike, say, study the pants like Manu did and then create a knock-off? Well, it's just not in Nike's DNA to sacrifice some profits to create a high-end pair of pants. The company makes good-quality clothing, but it's essentially a marketing machine. Besides, Lululemon invented a new category of workout clothes, in much the same way that Apple invented a new category of handheld devices with the iPhone. By moving first with a high-quality, well-designed product, Lululemon established the price of entry to the market and also established a strong, feel-good-do-good brand identity. That, married to its high-quality design, is a hard-to-beat combination. There's no gap in the yoga market that it's not filling, and its customers are fanatically loyal.

One way that Lululemon ensures loyalty is to target high-profile yoga studios in any new market and connect directly with people there, who become brand ambassadors. So by the time the company opens a store in a new location, it already has dozens of quasi-representatives in that neighbourhood. This grassroots approach is totally in keeping with Lululemon's brand, and it's been one way to downplay the inherent contradiction between yoga's spiritualism and the company's own pricy pants. So strong is the company's branding that even some really nasty PR moments have had negligible impact. First, its claim that some of its clothing was made with natural seaweed content was challenged by the *New York Times*. Despite Lululemon trying to prove otherwise, regulators in Canada insisted that health claims associated with seaweed be removed from the clothing. Even the murder of one sales associate by another in Bethesda, Maryland, didn't taint the brand (though the company can't have loved the press references to the "Lululemon Murder").

Of course, Lulu devotees also have nowhere else to go. Other athletic apparel companies try to get customers in touch with their inner jock. Lululemon's value proposition is about making your life better, healthier, longer—while making you more attractive, too. No one else was doing that when Chip Wilson entered the yoga gear market. That's still the case, and the reason he now owns it.

CHAPTER FIVE

Dream Big

Addison Lawrence is known as "Preacher" by his colleagues, who joke about passing a hat to take up a collection after his impassioned speeches at departmental meetings. Within five minutes of meeting him, you understand why. Dr. Lawrence's zeal for shrimp borders on the messianic.

A scientist whose expertise is shrimp and starfish aquaculture—the marine equivalent of agriculture—he has a CV that runs to seventy-three pages and includes more than 400 publications, with titles like "Classification and quantization of phospholipids and their dietary effects on lipid composition in shrimp." But Lawrence, who teaches at Texas A&M University, is one of those rare professors whose erudition doesn't get in the way of his enthusiasm. In a lecture hall packed with 200 students, he ambles out from behind the lectern and roams around, trying to make eye contact with each one. In conversation, even when explaining obscure scientific concepts, he doesn't expound so much as exclaim "Wow!" and "Jiminy crickets!" and "This will blow your mind!"

Shrimp, he tells me excitedly, are "just *loaded* with highly unsaturated fatty acids and the good cholesterol. And it's all protein!" Then, conspiratorially, "Want to dream with me for a minute? Today, shrimp is a luxury, but in twenty to thirty years, it will be as common as chicken. There could be a shrimp farm in every city in America." If he's right and shrimp is indeed the new chicken, Lawrence—a seventy-six-year-old with a Yosemite Sam accent and a faded Texas A&M University cap situated jauntily atop a thatch of tousled white hair—is the man who will have made that possible.

Americans already eat a lot of shrimp; Las Vegas casinos alone go through close to 70,000 pounds a year. In fact, the United States consumes more shrimp than any other country in the world, annually polishing off about $4 billion worth, 94 percent of which is imported, primarily from Asia. Shrimp account for 35 percent of all American seafood imports, by far the largest proportion, more than salmon, crab, tuna, scallops and squid combined. In other words, the country runs a large shrimp deficit in terms of balance of payments because, as shellfish go, they're expensive.

There are two reasons they are pricy. First, they're not so easy to catch in the wild (partly because there simply aren't enough of them), and second, to farm them, you need very warm water. Even so, there's a limit to what a shrimp farm can produce. Saltwater ponds yield up to 20,000 pounds annually per acre of water; about 50,000 pounds can be harvested annually per acre using raceway systems, where shrimp are raised in rectangular, trough-like containers that range from 300 to 2,000 feet long and 6 to 15 feet wide.

But because the growing season is year-round in the tropics, it's possible to harvest two and a half crops annually there, compared to a single harvest in the United States. Consequently, shrimp farming, once a licence to print money in places like Texas and northern Florida, is on the verge of extinction in America. "Today, shrimp farming in the U.S. is 25 percent of what it was in 2000," says Lawrence. "Our farmers are going bankrupt." And quickly.

Unless, that is, they get their hands on Dr. Lawrence's new system, in which case they can put more than a *million pounds* of shrimp on ice per acre per year—a more than twentyfold increase in production over current farming methods. What's his secret? Well, it's not a secret any more, as it's been patented and is in the process of being commercialized. But the innovation is, as he puts it, "simple as heck."

To come up with an idea that's both simple and revolutionary, however, he had to ask the right questions. The most important of which turned out to be, why wouldn't this work?

JUST IMAGINE

Little kids have big imaginations and few doubts. When a four-year-old tells you he's a superhero who's going to save the world, he really believes it. Innovators also set their sights high. They, too, believe they can change the world in some way. After all, why not?

But innovators are also pragmatic idealists who enjoy wrestling with tricky problems. Instead of seeing them as annoying obstacles blocking the path to fame and riches, as dreamers do, innovators generally seem to relish challenges and to have real staying power. Why is running SawStop Steve Gass's dream job? It's challenging and interesting. Why did the Project Darwin people throw themselves into the task and view it as a career highlight? It was challenging and interesting.

As I learned when I started working at the *Financial Post,* where I reported on technology, many innovators go into business not because they expect a giant payday but because they want the opportunity to follow their interests. They want to see how far they can go. Many do it to have the chance to build something from scratch, and because they believe in and are thrilled with the product, process or service they've created.

Since our ancestors first stood upright, people have been innovating: more and better tools, more effective and efficient ways of doing things. It seems to be a basic human impulse. And a lot of people do it because they believe that, in some small way, they can make the world a better place.

A BETTER WORLD, VIA CRUSTACEANS

The distinct twang of Missouri—"Missourah," as he and other natives say it—is still there in Lawrence's voice, and he's got the can-do attitude you'd expect of a man who, growing up, spent his summers wandering barefoot on the banks of the Mississippi and picking 200 pounds of cotton a day on his grandparents' farm. His beloved grandparents were the kind of people who shot squirrels for supper and didn't see any point in schooling girls past the eighth grade, but Lawrence's mother had a lucky break. When her little sister refused to go to school, period, she was deputized to drag her back and forth and thereby managed to get a high school diploma herself, and then a nursing degree. Lawrence's father, too, was a firm believer in education. The first in his family to make it past high school, he became a teacher and eventually the superintendent of schools in Cape Girardeau, Missouri.

Like all the other boys in town, Lawrence spent a lot of time playing basketball and baseball, and he was good—good enough to entertain dreams of going pro. But then he ran the numbers and realized that, most likely, he'd wind up the same way the majority of small town boys with big league dreams do: broke and on the sidelines. So, instead, he wound up an undergraduate at Southeast Missouri State University with four majors—chemistry, math, biology and education—plus a minor in physics. Always hungry for knowledge, he wanted to keep his options open. Maybe he'd be a teacher like his dad. Or maybe an engineer. Or a doctor.

A smart guy with a mind for detail, he could easily have taken a traditional route to money and security. But curiosity was the primary driver for Lawrence, and he had a passion for research and acquiring knowledge. Even at twenty—"When I thought I was *so* mature! That really gets me today," he chuckles—he knew he wanted to "break new ground." The reason was as much connected to self-actualization as to human progress. “When you stop learning, you’re dead,” Lawrence says. “As far as I’m concerned, what keeps you alive is the quest for new knowledge, continuously. Think about it. The more you know, the more you realize you don’t know.”

He just didn’t know how, exactly, he was going to become an intellectual pioneer. What route would he take? By the time he’d finished graduate school at the University of Missouri, where his doctoral dissertation in physiology involved a close examination of bullfrog guts, Lawrence had narrowed his options down to “the two big unknowns,” space and the ocean. Then the National Institutes of Health awarded him a two-year fellowship at Stanford, where he studied with one of the foremost marine and cell physiologists in the world. That helped him decide: the ocean was a heck of a lot more accessible than space in the 1950s, not to mention teeming with forms of life that no one knew too much about at that point.

There was plenty of scope for a scientist who wanted to break new ground. In California, he studied the sea urchin and abalone, but when his parents complained that he was living too far from home, he relocated to the Gulf coast of Texas, where he’s remained ever since. “I was shocked by the lack of diversity of marine life in the Gulf compared to California,” he remembers, but he soon recovered and decided to focus on shrimp, in part because no one had even figured out how they reproduced. “I lucked out,” Lawrence says. “That was the beginning of my dream.” He had a crazy idea: maybe one day it would be possible to farm shrimp commercially.

He rose rapidly through the ranks of academia while his wife, Jackie, had one baby after another. "It was her choice, not mine," he says. Not that he's complaining; he is one of those fathers who could happily talk your ear off on the subject of his children. But by the time their fourth son was born in December 1966, Jackie decided "Enough!" and started taking the Pill. She'd been instrumental in polishing Lawrence's rough edges; when they were dating, she'd taken him to fancy dinners at her college and taught him which fork to use. They'd married young, while he was still in graduate school and she had the kind of energy and spirit that's required to bring up a passel of boys. But one Sunday afternoon in August 1967, complaining of fatigue after they'd come home from church, she decided to take a nap while Lawrence mowed the lawn. As will often happen with young children, one of the boys barged into the bedroom shortly thereafter, but he couldn't rouse his mother. She was dead. Lawrence, still struggling with his emotions nearly half a century later, says, "I found out later that the Pill caused heart attacks in women who'd had rheumatic fever, as my wife did when she was young." He was left with four little boys and every reason to give up. "I would never have made it through that period if it hadn't been for my mother—she came down to help me."

And it was around this time that Lawrence began thinking about his purpose in life. What, at the end of the day, was the point of studying shrimp, aside from satisfying his own curiosity and putting a roof over his kids' heads? Fundamentally optimistic and interested in the world around him, he realized he just wanted to make a difference, in whatever small way. He was a person who cared, above all, about contributing to "the betterment of mankind." But how was he going to get there? How in heck could studying shrimp improve the lot of others?

It wasn't immediately clear. The recognition of his end goal, however, coincided with an explosion in scientific knowledge about shrimp. Researchers had figured out how shrimp reproduce, which helped make domestication possible. "Oh my heavens," Lawrence says by way of preparation. "I think their mode of reproduction is unique. The female emits a pheromone that attracts the male, who follows her through the water for a while, then turns upside down underneath her; both rotate crossways, and the male contracts his tail and grasper for about five to ten seconds as he deposits sperm in the female's thelycum. Eight to sixteen hours later, she spawns anywhere from 50,000 to 500,000 offspring." In order to mate, shrimp need a fair amount of room. "When I give my talks to Rotary Clubs I just tell 'em the male needs at least an eight-foot running start in a straight line." But because shrimp reproduce every eight to nine months in captivity, "within a year we have a second generation that has already been genetically selected and evaluated as to what we want. So every year we have a new commercial breeding population. To develop a new commercial herd of cattle would take you fifteen to twenty years."

By 1980, about 1 percent of all shrimp being consumed worldwide was raised on farms. By today's standards, they were rudimentary farms, employing what Lawrence refers to as "the rocking chair method": take a large pond, dump in some small shrimp you've managed to catch in the wild, sit back in your rocking chair on the porch and hope for the best—which was, at that time, between 500 and 1,000 pounds per acre of water per crop. But then he and others started wondering, what would happen if you fed the shrimp special food? Would they grow faster? And if so, what should be in that food? Answering these questions led to a new area of expertise for Lawrence: marine feed and nutrition.

Turned out that feeding the shrimp was a terrific idea. Yields doubled. But that created a new problem. At a certain point, the shrimp started to die off or stop growing. What was going on? The answer: if they got too big or if there were too many of them, they began starving for oxygen. Aerating shrimp ponds was the solution, and that bumped yields back up again, ushering in the golden age of American shrimp farming. With production levels of 5,000 to 10,000 pounds of shrimp per acre of water, a capital investment of just $10,000 to $30,000 per acre, and shrimp that had been genetically selected to grow more quickly, "farmers were laughing all the way to the bank. They were making a *lot* of money."

Money held no particular allure for Addison Lawrence, who by this point had remarried and had another child, a daughter. But he was beginning to see how his interest in shrimp could lead to the betterment of at least some portion of humankind. If he could help farmers grow even more shrimp, even more quickly, it would be good for them, good for the U.S. economy and good for shrimp eaters, too. A shrimp lover who tucks into crustaceans about once a week,

Lawrence has a soft spot for those of us stranded in northern climes. "If you're not on the Gulf coast or the southern Atlantic coast, you may really never have tasted the high-quality shrimp. Most of the shrimp that are imported, would you believe, have been frozen and thawed at least two times and maybe up to four times." If there were more shrimp farms, we, too, could eat them fresh.

By the late 1990s, American farmers needed someone like Lawrence on their side because Asian farmers were pushing them out of the market. The math was pretty simple. To remain viable, American shrimp farms had to do what the Asian farms were doing—grow shrimp every day of the year. But how on earth were they going to do that given the weather? It was the kind of challenge Lawrence had been looking for his whole life, complete with never-ending learning opportunities, a sense of urgency and the opportunity to do work that might make people's lives better in a very real, quantifiable way.

"I'm a bit of a Don Quixote," he acknowledges, and certainly there were plenty of windmills to tilt at, the first being the U.S. climate. The minimum water temperature for commercial breeding is twenty-five degrees centigrade, but thirty degrees or even higher is better. Maintaining that heat year-round would, even in the southern states, necessitate a greenhouse or some other sort of structure—which, right off the bat, would increase capital costs significantly.

Estimates of production potential suggested shrimp farms operating raceways year-round might produce 25,000 shrimp per acre per year, but "the accountants," as Lawrence calls them, announced that to justify the additional costs of indoor operation, the yield needed to be *ten times* that, 250,000 shrimp per acre, in order to break even. "'Oh my God' is exactly what we said," recalls Lawrence, a religious man whose speech isn't generally flavoured with such outbursts. "'Oh my *God*!'"

But the accountants weren't finished. More bad news: the shrimp also had to grow faster—not the one gram per week the farms were then getting, but one and a half to two grams per week. "And the other thing they told us was, 'Hey guys, you've got to lower your feed costs *and* use less feed per pound of shrimp that you produce.'"

It all may have sounded like an impossible list, but not to Lawrence, who relied on "that old American ingenuity," the kind he'd learned growing up in the Ozarks. He viewed the problems facing shrimp farmers as exciting opportunities to discover new things and "to go to places where others aren't going. And hey! This could better the standard of living for mankind at the same time." He just needed to figure out how to do it.

THE *F* WORD

Fear of failing is one of the biggest impediments to our natural desire to make things better and to innovate. Because wrong answers and mistakes are stigmatized in so many schools—and families—many of us try that much harder to locate the "right" answer as quickly as possible.

One reason innovative thinkers view challenges as positive opportunities rather than threatening tests is that they view failure much more benignly—it's one way to weed out answers and approaches that don't work. Failure, viewed through this lens, isn't catastrophic. It's just part of a natural process of elimination that clears the path to success. In this way, too, innovators share a habit of mind with really young kids, who are fearless about being wrong once in a while. Preschoolers are delighted when they figure something out, but not particularly upset—and rarely, if ever, embarrassed—when they make mistakes.

Innovation by definition involves blind alleys and wild goose chases. Take Matt Feaver's sea lion mission—what, really, did the Project Darwin team learn from his study of male versus female sea lions? If the measure of success is what his mission contributed to the end product, the answer is absolutely nothing, big waste of time—no gold star. But if the measure is whether his mission encouraged the team to think more innovatively, to take risks without worrying about making mistakes or looking foolish, the sea lion escapade would have to be rated a success.

In the one-right-answer world, mistakes are costly, embarrassing disasters on the road to ruin. But in the control-alt-delete world, mistakes are framed as growth opportunities. Recently, confessing to making lots of them has even become something of a badge of honour for business leaders. "I'm always telling people, 'Look, I make a mistake every day, but hopefully I'm not making the same mistake twice,'" Peter Löscher, CEO of Siemens, told the *New York Times* in 2011. "If you think that you're not making mistakes then you are not making the tough decisions that you should make as a leader."

Senior executives at DreamWorks Animation go even further, explicitly reframing mistakes as necessary steps in the creative process. It's not just talk. When, midway through production on the 2010 hit *How to Train Your Dragon,* the creative team decided the lead dragon was too small and timid looking, they were allowed to stop production on the film and go back to the drawing board. Stopping production is hugely expensive, and the team eventually came up with a complete overhaul—a fierce, intimidating dragon—that wound up changing other aspects of the movie. But heads didn't roll. Rather, DreamWorks's head of human resources now proudly tells this story as proof of the company's conviction that "it is critical to empower employees to take risks, move boundaries and test the limits of their imagination. Simply put, individuals must be allowed to fail in order to innovate."

"The path to truly new, never-been-done-before things always has failure along the way," Regina Dugan said in a TED talk in March 2012. Dugan's job is all about creativity: she directs the Defense Advanced Research Projects Agency, the innovation engine of the U.S. Department of Defense. She showed the TED audience a video clip of a new, tiny, camera-bearing drone that can fly like a hummingbird—all directions, even backwards, and it can also hover and rotate—though it does not, Dugan noted, consume nectar. She proudly pointed out, "In 2008, it flew for a whopping twenty seconds. A year later, two minutes, then six, eventually eleven. Many prototypes crashed—many. But there's no way to learn to fly like a hummingbird . . . Failure is part of creating new and amazing things. We cannot both fear failure and make amazing new things."

However, as Harvard business professor Amy Edmondson has pointed out, failure promotes success only if you actually take the time to analyze your mistakes, whether they were made in a business start-up or in a relationship. Failure, she adds, has to be separated from fault, and for many people that requires a bit of deprogramming, as we learn early on that they are one and the same. To that end, Edmondson has devised a "spectrum of reasons for failure," which lists causes ranging from deliberate deviation (blameworthy) to thoughtful experimentation (not blameworthy).

In this framework, intention is extremely important. Inattention, for instance, is often classified as blameworthy—and if it is the result of lack of effort or lack of caring, that may well be the truth. But what if inattention is caused by fatigue? A tired child can't focus in class, and a tired shift worker will not perform optimally at the end of the shift. Is the child or the worker to blame, or is there a systemic problem, like needlessly long classes or endless stretches between breaks?

Edmondson notes that when she asks executives to classify their organizations' internal business failures using her spectrum, the execs usually conclude that only 2 to 5 percent of the mistakes made are actually blameworthy. However, they report, 70 to 90 percent are treated as though they *are* blameworthy errors. One side effect of this cause-and-effect gap is that failures go unreported, hidden away as shameful events. And then, of course, no lessons at all are learned.

THE LONG SLOG TO EUREKA

Across the United States, Addison Lawrence and other researchers began working on the interrelated set of shrimp farming challenges: grow more, more quickly, year-round, with less food. The rest of them immediately zeroed in on greenhouses, because they're cheaper to build and operate than structures that need to be heated.

Lawrence, however, didn't rush to any conclusions. He took a big step back and questioned the problem. Was the real goal to save farmers who were already in the shrimp business? No, he decided. That would be nice, of course, but his *real* goal was to enable even more farmers to raise shrimp. And the reason for that wasn't just to help the U.S. economy, but also to feed the world. "With the globe's population set to double in the next twenty to fifty years, where is the extra food going to come from?" is what he was wondering. "On land, food production is growing at only 1.5 percent a year. And we can't get more fish out of the ocean—our natural fisheries are at maximum sustainable yield. The farming of aquatic animals is really the only answer."

For him, then, the real question was how to grow enough shrimp to feed the world. When he dreamed big and thought about it that way, the greenhouse model just didn't make sense. It was too dependent on sunshine, thereby dramatically limiting possible locations of shrimp farms. To feed the world, you've got to be able to raise shrimp just about anywhere, which means you need heated buildings—which also means, however, that you need to find a way to get dramatically higher yields in order to justify higher capital costs.

"Okay," Lawrence thought, then mentally pressed control-alt-delete on the prevailing raceway model—flat troughs stretching hundreds of feet—which he'd helped to develop. Starting from scratch, he asked himself, what *would* be a good way to grow shrimp indoors?

And then came the connection that may have made all the difference. Some colleagues over in the Agriculture Department were developing a vertical farming concept—literally, high-rises where crops would be grown on every floor. He soon decided that vertical might be the solution for shrimp, too. He could picture it in his head perfectly: row upon row of containers full of nice, plump shrimp, stacked to the ceiling. But when he went to build a prototype, disaster. When filled with three feet of water, the containers were just too heavy to stack.

Lawrence realized he was dealing with an assumption that shrimp need three feet of water, because that's how it's always been done. Again, he pressed control-alt-delete and decided to work backwards. "I went up to the engineering school and said, 'Hey guys, what's the maximum water depth that I can put in a raceway and economically stack 'em?' The answer was twelve inches, maximum. I said, 'Oh my gosh!' The typical *minimum* at that time was twenty-four inches."

All right, he decided, let's see if shrimp can survive and grow in twelve inches of water. Could you achieve commercial production levels with such shallow water? Would the shrimp still have enough room to perform their weird mating ritual? Lawrence conducted his first experiment in early 2002, and when the results came in, he says, "Man, was I shocked! The very first experiment I did was so successful, I couldn't believe it. So here I had a situation where, doggone it, conception-wise, it looked like I *could* stack the raceways, and have them four, fix, six, seven raceways high, and justify the cost of my building."

The prototypes, built in a darkened room a few feet from Lawrence's office at A&M, had raceways stacked four high. On the top level, baby shrimp, who moved down through the stack as they grew. Lawrence, wearing khaki shorts and a faded Hawaiian shirt, was in there checking on them all the time, tweaking this and that, aided by a handful of graduate students and lab technicians. This went on for five years while he continually improved the raceways. He designed a new way to keep the water circulating by simply raising the centre of the trough—before, they'd always been flat or bowl-shaped—and developed a new food, which is also now being patented. The long hours didn't bother him in the slightest. "I can outwork an individual twenty years younger than I am," he boasts, and the reason is intellectual curiosity. "If you really enjoy your work, you can work sixty to eighty hours a week and enjoy every moment. If you don't enjoy it, you'll force yourself to work forty hours and you won't be happy."

In the end, his stacked system was able to produce somewhere between 600,000 and 1.2 million pounds per acre of water per year. Not only that, but the shrimp grew 40 percent faster because over the years he had figured out how to manipulate the water's temperature, current, salinity and alkalinity to promote maximum growth, all of which is possible only in a facility with a constant temperature. A building, in other words, not a greenhouse. He didn't just meet the milestones the accountants had established. He surpassed each and every one, largely because he'd reframed the questions and goals and was taking his measure with a different yardstick: How can I help solve the problem of world hunger?

Other research teams across the country made progress, too, some achieving shrimp yields as high as 300,000, but no one even came close to Lawrence's results. But his key breakthroughs, as he's the first to acknowledge, were both very simple conceptually: shallow water, stacked tubs.

By 2007, Lawrence knew he had what he had been aiming for, and A&M encouraged him to apply for a patent on the raceways. Meanwhile, the university, which controls rights to the technology, since it makes Lawrence's research possible, licensed it to a commercial firm, Royal Caridea, which will begin production in the next year or so; Lawrence will get a share of the royalties after the university takes its cut. Lawrence unveiled his design at an annual conference of marine and aquatic experts held in early 2011 in landlocked Las Vegas and promptly became something of a celebrity in the world of aquaculture: he made the cover of *Fish Farming News,* the grin on his face and twinkle in his eye making him look like a slimmed-down Kris Kringle.

There are still teams of researchers investigating the use of flat raceways, but the baseball player in Lawrence's heart isn't worried. "It remains to be seen is who is gonna score, but I wouldn't bet against me," he says. "What really excites me is that I think this is just the tip of the iceberg." His confidence, he says with a laugh, comes from years of missteps and miscalculations. “The reason I’m doing the best work of my life right now is because I have fifty years of mistakes telling me what to do.” Lawrence looks at his life pretty much the way he looks at shrimp farming: there’s always room for improvement, always something new to figure out. “I’m learning to interact with other people better and how to better express myself,” he says thoughtfully. “Also how to enjoy and be happy with other people. I’m better at that today than I was ten years ago. I think I’m still maturing.”

The kid from the Ozarks who didn’t travel more than 150 miles from home in his first seventeen years now circles the globe a few times every year, and has been entrusted with more than $16 million in research grants. He still works seventy hours a week—“I’ll die with my boots on!”—and is enthralled with a new big idea cooked up with his brother, a biologist who specializes in starfish. The long and the short of it is that, based on their research involving the starfish, which shares 80 percent of our DNA, they believe they’ve come up with a way to produce stem cells for medical research. “Do you want to dream with me for a bit?” Lawrence asks again, but I know by now that it’s not really a question, and in any case, he’s not a guy who takes no for an answer. “This could be a source of stem cells for the regeneration of human tissue. That’s a biggie, isn’t it? That would be . . . wow!”

Lawrence knows this idea is still a little *too* innovative for the mainstream, but he’s used to that. He’ll just keep on asking questions and dreaming crazy dreams with his wife, Rose, as they sip their coffee in the morning on the veranda of their home on the shores of the Gulf of Mexico, watching the sun rise and waiting for the world to catch up. So far, it always has, sooner or later.

PERMISSION TO DREAM IS ALSO PERMISSION TO FAIL

One thing that's really striking about the innovators I've met is their lack of cynicism. Some are dry and sharp—you probably need to have a sense of humour in order to stick it out as an innovator—but they all retain what I'd characterize as a childlike sense of optimism. That's not an insult, but a compliment; somehow, they've managed to hang on to a sense of wonder about the world and a belief in their own ability to have an effect on it. They're persistent and stubborn about reaching their goals, but flexible and open-minded when it comes to figuring out how they're going to get there. They're enthusiastic and engaged, always searching for a better way, and then shrugging and moving on when they fall short.

That's not just a good way to run an aquaculture laboratory or a business. It's also a good way to run your life: allow yourself to dream big, forgive yourself when you don't quite get there and then try another path.

Many people, however, fear that dreaming big will lead to disappointment and conclude after a few mistakes that they're simply not up to whatever new challenge they're facing. Unfortunately, this can become a self-fulfilling prophecy because shame doesn't encourage innovative thinking. Quite the opposite. Researchers have shown that self-censure is "inimical to original creative thinking." In other words, beat yourself up and your capacity for innovative thought evaporates. "Self-compassion"—taking a non-judgmental attitude toward your inadequacies and failures, as Addison Lawrence does—is what promotes creativity. If you can forgive yourself for mistakes, you are not only more likely to learn from them, as Amy Edmondson points out, but you are also more likely to think innovatively.

In one study, participants were told to write a detailed account of something that happened in high school or college that made them feel really bad about themselves, "something that involved failure, humiliation or rejection." The control group wrote only about this horrible experience, but the second group was given a few minutes to perform three additional writing tasks: list ways that others experience similar events, write a paragraph expressing kindness to themselves in the same way they would to a friend who'd had a similar experience, and try to describe the event in an objective manner. These prompts were all aimed to induce self-compassion by getting subjects to reframe their bad experiences as things that could happen to anyone, rather than as proof that they were awful or inadequate human beings.

Next, both groups took a classic test of creativity. The control group, the ones who'd basically been castigating themselves in writing throughout the experiment, scored poorly compared to the self-compassionate group. The people who'd given themselves a break, and perhaps even a pat on the back, came up with more creative and original answers. Maybe that's because they'd given themselves permission to make mistakes.

One thing is for sure: without self-compassion, it's much more difficult to move beyond assumptions and think innovatively about problems. It feels risky and uncomfortable to press control-alt-delete because it's almost certain you're going to make some errors, and that would be embarrassing. But paradoxically, the less willing you are to make mistakes, the more likely you may be to make them because you've narrowed your mind and drastically reduced your openness to new opportunities.

This is one mistake innovators don't make.

CHAPTER SIX

Borrow, Don't Just Follow

In Soviet Russia in the 1940s, a twenty-year-old naval patent agent named Genrich Altshuller began reviewing existing patents in various fields, looking for common denominators. Did innovative ideas come out of the blue, or was there some discernible pattern of invention?

After analyzing 40,000 patents, he recognized that there was a pattern, and established an important rule: An invention is the removal of technical contradictions. For instance, to make a car go faster you need a larger engine—but then the car will be heavier and therefore slower. Removing the contradictions, rather than accepting negative trade-offs, requires invention. You have to create a lighter yet more powerful engine, or find other ways to reduce the weight of the car. By categorizing patents, Altshuller recognized that certain types of contradictions repeat so reliably across industries and sciences that he was able to come up with a matrix of common contradictions.

Another consistent pattern: innovations that remove those contradictions rely on scientific effects outside the field in which they were developed.

In other words, if you're wrestling with a problem, chances are good that someone, somewhere, in a different field has already wrestled with the same sort of contradiction—and the most efficient and effective way to solve your problem is to imitate some aspects of their solution.

We've already seen the truth of this: Addison Lawrence resolved his contradiction—the need to breed more shrimp versus the expense of flat breeding tanks—by borrowing the idea of stacking from agriculture. Similarly, at SawStop, to resolve the contradiction of a dangerous tool versus user safety, Steve Gass borrowed the type of sensor that had been developed for touch screens.

Altshuller's theory of inventive problem solving, which is known by its Russian acronym TRIZ, helps engineers break down the process of problem solving and innovation into discrete chunks, using, among other things, the matrix of common contradictions, forty principles of invention and an algorithm for innovation that's continuously being refined by TRIZ scholars and devotees worldwide. There are many of them, because the TRIZ methodology has been used successfully at places like Boeing, Hewlett-Packard, Johnson & Johnson and NASA. TRIZ involves some pretty complex science, but the core insight is one anyone can grasp: Why reinvent the wheel if you can borrow and adapt one from another field?

That Altshuller endorsed a multidisciplinary approach makes sense given his history. In 1949, he and a colleague apparently fired off a letter to Stalin suggesting a few ways that inventive problem solving could get the country back on track after World War II. Stalin didn't take too kindly to the implied criticism, and it didn't help any that Altshuller was Jewish. Interrogated and tortured until he "confessed" to being a dissident, he was then packed off to Vorkuta, a labour camp located 100 miles above the Arctic Circle. Imprisoned in the gulag along with some of the leading lights of the intelligentsia, Altshuller decided to make the best of it. At night in the barracks, he asked others to teach him what they knew; physics, literature—it didn't matter, he just wanted to learn. On some level, he knew that curiosity improved his chances of survival. So long as he remained interested in life and felt he had things to learn, he had a strong psychological motivation to keep going in brutal conditions that were killing many other prisoners. And one thing he learned was that ideas from one discipline could shed light in another.

"This camp was first a place of education. He studied fourteen, sixteen hours a day," Boris Zlotin, Altshuller's long-time colleague, told a journalist. Once, surprised by Altshuller's encyclopedic knowledge of Verdi, Zlotin asked, "How you do know these? You had time to go to the opera?" Altshuller replied, "Never, but my neighbour in the barracks was the world's best specialist on Verdi's music, and he would sing me all his operas at night."

After Stalin's death in 1953, Altshuller was freed from the prison camp and continued his research and work on TRIZ—while cranking out several science fiction novels under the pen name Genrich Altov. He viewed his writing as helpful to his research, and once observed that science fiction "helps overcome psychological barriers on the road to [the] 'crazy' ideas without which science cannot continue its development."

Altshuller died in 1998, but TRIZ is still going strong. Today, databases used by computer-based TRIZ programs include insights generated by reviewing millions of patents, and the applications of knowledge from one industry to another continue to expand, too. For instance, when the cost of energy went up and some dairy farm operators could no longer afford to dry cow manure using now-expensive heat, they discovered, via the TRIZ methodology, that juice-makers had dealt with the same contradiction and come up with a way to concentrate fruit juice that required no heat—a solution that dairy farmers could easily borrow for their own purposes.

LOOK FOR ANSWERS IN UNLIKELY PLACES

Rolf Smith is big on borrowing because he knows it can kick-start the process of innovation. He remembers one thinking expedition he led for General Mills, where the goal was to increase the sale of cereal by 6 percent by changing packaging. Smith packed colleagues off to the mall in carefully selected teams of three that included an adaptive person who liked to work within existing systems, an originator type who preferred to make big changes to the systems themselves, and an in-between person who bridged between the other two. The assignment: Go into stores you've never been in before and find out how they're packaging stuff.

One big idea came from RadioShack, which packages televisions simply by displaying them side by side playing different sporting events, so customers stop to watch. Another idea came from a pet store; no one in one group had ever shopped for a pet, so they didn't know how they were sold. One guy picked up a puppy, held it for a moment and recognized that the lure was simply to try to get people to pick up a dog—once they had, it was very difficult to put it back in a cage and walk out of the place empty-handed. The third big idea came from some shiny, reflecting signage inside the mall, which caught one team's collective eye.

What these three seemingly random ideas translated into was a new Wheaties box featuring an eye-catching shiny hologram (flashy, like the reflecting signs) of various sports figures kicking a football, say, or swinging a bat (different events, like the TVs in the store). When you picked up the box (like the puppy in the pet shop) and turned it, the image moved and became an action figure (TVs again)—the box was not going back on the shelf.

Later on, when Smith got a call from General Mills saying the cereal boxes were in stores, he drove to the supermarket with two of his daughters and a neighbour's son, and said, "Okay, one at a time, go down the cereal aisle and pick a box. I'll wait at the end and buy it for you." All three turned up with boxes of Wheaties. As they were leaving the store, Smith turned to the boy and said, "Robert, I guess you like Wheaties." Robert's reply: "What are Wheaties?"

"That hologram was a really successful innovation,"

Smith says today, "and sales increased by more than 7 percent." He knew, from his own little experiment, that it didn't have a lot to do with the product inside the box. It was all about the box itself, an inspired mash-up of borrowed ideas which, combined, equalled an innovation.

GOOD IDEAS CAN BE FOUND ANYWHERE ON THE FOOD CHAIN

Isadore "Issy" Sharp, the founder of the Four Seasons hotel chain, believes innovation centres on taking what already exists, then pushing the boundaries to see how much better you can make it. "The hotel industry is 500 years old," he points out. "There's nothing new. People come, they sleep, they eat. All I did was make that experience better."

Today, the preternaturally youthful eighty-one-year-old Sharp could spend leisurely mornings in his sunlit living room—a vast expanse of flagstone floor and plate glass window reminiscent of one of his hotels—but that's just not his way. He's still very hands-on with the chain he built from one motor hotel in Toronto to eighty-five luxury hotels around the world. Sharp is thoughtful in his manner of speech but just as passionate about the idea behind the brand as he was the day he began.

When he got into the hotel business five decades ago, his inspiration wasn't other high-end hotels—the son of an industrious but small-time builder in Toronto, Sharp hadn't stayed in any. He'd just noticed one of those invisible problems he thought he could fix: the discomfort factor of staying in a hotel where you felt like a temporary, anonymous inhabitant of room 1703. The crummy bedspread, the room service carts clattering down the corridor at 3 a.m.—these were just facts of life when you travelled. The experience was about dislocation, not comfort.

Sharp's big idea was that people should feel like honoured guests when they stayed in hotels. He started innovating in the bathroom, finding the best shower heads, stocking thick, cushy towels—the things he'd want in his own home. Four Seasons was the first chain to supply tiny bottles of shampoo and conditioner, the first to put bathrobes in the rooms, the first to invest seriously in the quality of its mattresses. But all those innovations could be—and have been—replicated.

The innovation that couldn't be copied was the unique culture of service the chain offered. One of Sharp's key insights about service came after opening Inn on the Park in London, England. It became the top-rated hotel there in its very first year, despite the presence of Claridge's and The Connaught, and Sharp knew exactly why. "We brought a different quality of service, rather than the formality of all those very famous and important hotels. If you were important, they bowed and scraped, but you couldn't walk in the way I'm dressed and get service," he says, though he's the epitome of elegance in a soft, cotton collared shirt and brown twill pants. "You'd be looked down upon." His service innovation: to treat *anyone* who walks in the door as a guest whose comfort is of paramount importance. It turns out what travellers, even high and mighty ones, really want is to feel welcome.

This idea about service didn't come to Issy Sharp out of the blue. He pinched it, actually, from another business—McDonald's. He admired McDonald's consistency and uniformity of product, which was seen as its great strength. But what really struck him was the attitude of its staff at that time. Teenagers who had no earthly business being happy or feeling pride of ownership in the french fries and burgers displayed exactly those qualities, and they made every customer feel welcome. And Sharp knew that if his own employees could do the same, his hotels would be successful.

But when Sharp instructed the senior managers at his growing hotel business to attend a training session at McDonald's, some of them thought he was out of his mind. "They were laughing about it, making fun. 'Here we are sitting in McDonald's, and we're the Four Seasons!' I saw that these people just didn't understand. And there was just no sense in trying to talk to them, it was almost a different language. That was a watershed for me. I believed that this idea would work, because why wouldn't it? Wouldn't *you* like to be treated properly?" Most of the resistant managers wound up leaving the company shortly thereafter, freeing Sharp to implement the company's central code—also borrowed, it turns out. It's actually the golden rule: Do unto others as you would have done unto you.

To innovate, Sharp had to identify several different yet connected problems: first, the discomfort of staying in hotels; then, the snobbery and indifference of staff at five-star hotels. He dealt with the latter problem by making sure every Four Seasons employee understood that he or she had a mandate to take action and solve customers' problems. Each and every employee has personal responsibility for pleasing the customer. Even a maintenance worker who might never interact with a hotel guest face to face knows how important it is to clean the windows properly, since they *will* be in front of a guest.

Unlike McDonald's, Four Seasons hotels aren't completely consistent—the one in Maui is significantly different from the one in New York. But the consistency of service never wavers, because everyone involved knows that they're in the business of making people feel important and valued, no matter what they're wearing or what titles are on their business cards. A one-star idea adapted to a five-star setting: it's been a winning combination.

IF YOU'RE GOING TO FOLLOW, KNOW WHERE YOU WANT TO END UP

"Where do I want to end up?" might seem like the most obvious question in the world, the one you'd need to ponder in order to decide where to go to school, whom to marry, whether to have kids—any major life decision, essentially. And yet sometimes, people simply don't think much about long-term goals, except in the vague sense of wanting to be wealthy or wanting to have a happy family. They surrender themselves to fate and drift along without a clear idea where they're heading.

Other times, people are very clear on their destination—they've seen others heading there, so it must be a good way to go. The beauty of following the crowd is that you don't have to ask yourself tough questions like, Why do I want to go *there*? Is it really right for me, given my strengths and weaknesses? Or have I headed in this direction because I think I ought to want to go there? Am I following others, or charting my own course?

That same urge to find a quick solution first and ask questions later has tripped up many a company, too. Never is it more dangerous than when technology is changing rapidly and the old guard, unable to fully understand the nature of the challenge, takes precipitous action. In the late 1990s, for instance, just about anything with *dot-com* in its name was white-hot, while old-media companies were increasingly perceived as yesterday's news. Because they were desperate to stay *au courant* but kind of clueless about how to do that, some crazy deals went down. The craziest of all was when an Internet service provider with wildly overvalued stock purchased a traditional media empire that had honest-to-goodness cash in its coffers.

The ill-starred relationship began, or so legend has it, with a chance meeting between two American executives in China. Gerald Levin, head of media conglomerate Time Warner, instantly liked Steve Case, who was at the helm of AOL. Seated close together at a government dinner, they got to chatting; Levin recalls thinking Case and his wife had a "cute" relationship, and AOL was, at the time, an Internet high flyer. Thanks to the dot-com frenzy, the ISP was able to use its stock—which had value on paper, to be sure, but that value depended very heavily on market sentiment—to buy a well-established media company that had grown slowly and surely through acquisitions, divesture and prudent management.

What was in it for Time Warner? Well, the sky was falling! Old media companies were dinosaurs, in imminent danger of being wiped out by the likes of AOL. Given that understanding of the problem and the threat, it made perfect sense for Time Warner to tie its content generation to the bright digital distribution prospects of AOL. It knew where it wanted to go. It needed to get in on this newfangled Internet thingy. Somehow.

If the executives—and the bankers who quickly lined up on both sides of the enormous transaction—had mentally pressed control-alt-delete on their own assumptions and stepped back to consider the nature of the problem rather than rushing headlong to find a solution that matched their perception, the deal might never have happened. In 1999, however, caught up in the excitement of what appeared to be a transformational moment in history, the market was nuts. Suddenly, businesses that had what seemed like good ideas were valued in the stratosphere, before they had even earned a dollar in sales, let alone profits. Concepts that had been used to establish the value of old-fashioned businesses were tossed out the window, and investment analysts rushed to find ways to justify the worth of new virtual entities. And thus, AOL, one of the first companies to introduce regular folks to the power and fun of the Internet, was worth billions—on paper, anyway. In fact, it was "worth" twice as much as Time Warner, although its business generated less than half the cash flow.

Time Warner was a company that knew how to move into new markets. Having begun life as Warner Communications, with music and movie divisions, as well as *Mad* magazine and Garden State National Bank, Warner had extensive experience acquiring and managing new businesses. For a time, the company owned Atari, which, as anyone of a certain age knows, was one of the first and biggest movers in online gaming and computing. The company also owned a number of cable channels, including MTV, Nickelodeon and The Movie Channel—all of which are still in existence, it should be noted. In 1989, Time Inc. made a play for Warner, largely to avoid being taken over itself; Paramount had made a $12 billion hostile move on Time, and one way to thwart that takeover was for Time to make its own big move.

The following year, Time Warner came into existence, and some aspects of that merger would be echoed eerily a decade later with AOL. For instance, because Time was the weaker partner—it needed the transaction more—the deal was tilted in Warner's favour. Warner shareholders were paid cash, which meant that Time shareholders had the less than pleasant experience of seeing the value of their stock decline. Henry Luce III, the son of Time Inc.'s founder, remarked, "Because of that son of a bitch at Paramount, we had to acquire Warner in cash. That made all of the Warner people rich and all the Time people resentful." It took almost eight years for Time Warner's newly combined stock to rise above the $200-a-share offer Luce had turned down from Paramount.

Nevertheless, the businesses made sense together, and after purchasing Turner Broadcasting in 1990, Time Warner became one of the world's dominant players in cable broadcasting. By 1999, the company was generating robust sales but also growing increasingly worried about how to stay relevant in the dot-com era. Would advertising dry up because of this new content delivery model? Insecure about the shift from analogue to digital, Levin, like media executives everywhere, was trying to get a grip on what the future held. Preoccupied with finding a solution ASAP, Levin didn't ask some other, highly relevant questions: Why is a company with lower sales and profit "worth" twice as much as we are? Why should we effectively sell ourselves to this company? Why do we need AOL?

The merger of the two giants that was announced in January 2000 was the biggest in history, creating a combined entity worth $350 billion. Unfortunately for all concerned, the dot-com era had peaked. Two months later, investors began stampeding out of Internet stocks, and by 2001, AOL Time Warner's combined market value had slid to just north of $200 billion. By 2009, when the romance was long over and AOL was spun off, Time Warner's value had dwindled to $65 billion. The company had to take the biggest loss in corporate history.

The writing was on the wall almost from the beginning. Time Warner's leadership really had no clear idea where they wanted to end up, or why. They were primarily following a market sentiment that the Internet was the place to be, and blindly trusting that their new partner was on the cutting edge and knew what to do. But AOL did not. Consequently, the partnership didn't result in much innovation, unless you consider failing on a massive scale and driving your shareholders wild with rage to be innovative. If you don't know where you're going, much less where to look for innovation, it's pretty unlikely that joining forces with others who have quite different interests is going to work out well.

One thing that fascinates me about the relationship between innovation and borrowing (versus following) is that, in one sense, this is stuff we all already know. It's what we teach our kids: of course it's okay to imitate—otherwise, how could you ever learn how to write, or do math, or ride a bike? You have to copy others to some extent to figure out how to do something new. But follow along blindly and you're likely to wind up in some kind of trouble. "Think for yourself, don't just do whatever your friend does," we instruct our children. "If Joey jumped off a bridge, would you jump off a bridge, too?"

In our own lives, we'd do well to remember this advice. We can successfully borrow to innovate, and sometimes, that means borrowing from ourselves—a solution that's worked to manage conflict in the office, for instance, might also work with warring siblings at home. One friend of mine with teenagers says that he consciously reminds himself, when discussing a heated topic, to use the same tone of voice that he uses with colleagues. "Heated discussions used to be the norm," he remembers. "There was a lot of shouting and sarcasm, and then slamming doors and sulking. And I'm just talking about me! After the millionth episode of this I realized I had to try something else, and I'm pretty good at defusing conflict at work, so decided to go with what works for me there. It's been one of those little changes that makes a big difference. We still get angry and there are still big disagreements, but things don't go nuclear as frequently and we're much more likely to get to a solution we can all live with around curfews and consequences and things like that."

Following the advice of others may also help you move forward, but only if you really understand the nature of the problem you're trying to solve and why this approach is the right one for you. Following blindly and trusting that what's worked for others will work for you is how Bernie Madoff's investors wound up losing their life savings and how AOL and Time Warner ended up in the equivalent of a bad marriage.

The fine line between judicious borrowing and blind following can be discerned with just a few questions: What is my problem, exactly, and where do I want to end up? Why did this approach work for others? Could I adapt it to my own situation? Am I doing this because I think it's a really good idea for me, or so that I can steam ahead without having to think any more? Answering those questions can help prevent a gigantic strategic error, like the AOL Time Warner merger, or a massive blow-up with your teenaged son. Once again, being willing to ask the right questions can open your eyes to new possibilities that will help you move forward, whatever your ultimate destination is.

CHAPTER SEVEN

Be Prepared to Change Course—Frequently

Innovation is rarely a linear path, even when the improvements that are being made are relatively simple and the result is clearly better than what already exists. There are almost always obstacles, not only to innovation but especially to implementation, and getting around them requires not just ingenuity but a combination of stubbornness and flexibility.

Just ask Sean Moore. In 1996, he was a fifty-year-old lawyer/commercial real estate agent/telecom entrepreneur in Houston making a decent enough living, although his ship had definitely not come in yet. Over beers one day, a friend in the real estate business mentioned a problem. Oval bathtubs were then all the rage—a selling feature in apartments, even—but they presented a real challenge because invariably, when you turned on the shower, water wound up on the floor. Shower curtains never closed properly at either end of the tub because they hung on straight rods.

Moore and his buddy, Dick Wise, batted ideas back and forth and arrived pretty quickly at what seemed to be the obvious solution: curve the rod so it paralleled the curve of the tub. Easy.

Except it wasn't. "It was really a geometry problem. And one reason I had gone into the law was to avoid math!" says Moore, laughing. "If someone had said, 'Look, I have this geometry problem, can you help?' I would have said, 'Forget it.' But since I didn't realize it was a geometry problem, I kept at it." He notes ruefully that many a tenth grader who doesn't see the point of geometry might have a change of heart hearing how it made him a millionaire, and then some.

But back then, of course, he didn't have any inkling that that's how the story would end. He just thought he was trying to solve a fairly isolated problem—alone, because Wise soon drifted back to his full-time job. Moore, however, was captivated by curved shower-curtain-rod dilemmas. It was amazing to him that something so conceptually simple could be so difficult to bring into existence. How, for instance, could you securely attach a curved thing to the wall? Through trial and error, he figured out it was possible with a "tangent," a straight extension attached to the curved part of the rod. After a year of wrestling with such issues, he had a prototype and was ready to test what he'd built. He called a friend who ran a supply company to see if he could borrow an oval tub, but the guy didn't have one in stock. Well, that was a disappointment.

But Moore, eager to see how his creation worked, decided to frame it up with a regular rectangular tub and just use his imagination. Stepping back to admire his handiwork, he had an epiphany. The curved rod would work even *better* with a standard-issue tub! "Think of all those shower curtains bunched up at both ends of the tub, pasted to the wall to prevent leaks, and all those people standing in the shower at a slight angle so the curtain won't touch them." The curved rod pulled the curtain out and away from the centre of the tub, so there would be no need to shy away from a slightly slimy shower curtain—and yet it took up no more space in the bathroom, as it was mounted high up on the wall. It was an elegant solution to a problem no one had ever clearly articulated, not even Moore, who'd been pondering shower curtain rods for months.

He saw that he'd got the right answer to the wrong question. The problem wasn't that trendy oval tubs needed a new kind of rod. The real problem was that showering in a traditional rectangular tub, trying to avoid contact with a gross shower curtain, was an unpleasant and uncomfortable experience. The status quo stank. But everyone was so accustomed to it that no one had thought to ask, Can I make it better?

Moore knew that he'd found the way to do that and promptly forgot all about oval tubs, telling himself, "This is it! This is what I can sell!" The curved rod seemed like a slam dunk. Its appeal could be summed up in just six words: "Stops leaks, adds space, looks great." So in 1997 he incorporated as Crescent Rod, but then "wore out a pair of knee pads begging for financing." He didn't get any. So Moore remortgaged his paid-off house, borrowed from friends and family and kept at it. Finding someone to license and manufacture the rod was just another problem to be solved.

Although money was tight, when Moore eventually landed a meeting with the CEO of a plumbing fixture firm, he gambled on a good outcome. Sure, the meeting was going to cost something—he had to get a plane ticket, get a hotel, rent a car—but this guy was a major manufacturer of straight rods. Moore was ready to break out the champagne. A licensing deal was guaranteed, he figured, because his innovation was so clearly superior to anything on the market.

However, the CEO didn't turn up for the meeting on time. He kept Moore waiting. And waiting. And though persistence is Moore's strong suit, being stood up really irks him, so the longer he waited, the more irritated he became. Finally, the CEO sent out his head engineer to schmooze Moore while he waited, but Moore was not in the mood. "I wouldn't say I was rude, but I wasn't chatty, that's for sure." So the engineer did all the talking, and the only topic of conversation was how misguided Moore's concept was and why it would never ever work. The engineer assured Moore there was no way he'd get a patent on a curved rod because there was simply no way to manufacture a curved rod that could telescope (contract for packaging and shipping purposes, then expand to fit between two walls). But instead of being disheartened, Moore got more and more excited, though he tried not to let it show. He was thinking, "This is great! This is my main competition and their top engineer is telling them it can't be done. But since I had no idea it couldn't be done, I've already done it." Instead of spilling the beans that he'd already achieved the impossible—as he surely would have told the CEO if he'd shown up on time—

Moore excused himself and headed home, considering the money he'd spent on the trip an excellent investment.

He realized he needed to change course. The quest wasn't to find someone else to take his product to market; the quest was to figure out how to take it to market himself. Unlike Steve Gass at SawStop, who found partners to help him do that, Moore decided to go it alone. So he doubled down, withdrawing money from insurance policies and getting a $100,000 line of credit from the bank. All the while his wife, Judith, held her tongue, which he viewed as the best possible form of support. Well, that and the hands-on help she gave later, packing rods into the back of a truck.

If Moore had gone to business school, he would have learned that it doesn't take something high-tech to change an industry standard; often an incremental improvement will do the trick. They might have taught him that the best products are those that solve everyday problems people don't even know they have. And they surely would have said that being able to explain the product's value in just a few words was golden. Well, Moore had all that going for him, but he still couldn't land a big sale. He knew that when people saw the rod they understood immediately, as he had, why it made sense. "I could put it in someone's house for three weeks and then ask for it back, and I knew they'd pay whatever I wanted to keep it." But house by house is hardly the way to get a business off the ground.

So Moore found out about trade shows, and by making the rounds at those, he got to know folks at all the big bathroom fixture firms. But even after he was on a first-name basis with tub-makers and rod-makers, nobody would buy his product. And while he was fighting to keep his innovation alive, Moore continued to spot new problems with it. Whereas problems might have been the final disheartening straw to another person, Moore saw them as opportunities, in a sense, answers to a simple question he kept asking himself: How can I make this thing even better? So he improved the fixtures that secured the rod to the wall and found even better plating for the rod itself. And because it's difficult and costly to ship a six-foot-long rod, he figured out how to make the thing in two pieces that would then telescope out to full length—a feat the engineer at the plumbing fixture company had told him was impossible. These kinds of incremental improvements along the way meant that when he did start selling the rods, they were good ones that customers really liked.

His first break came when a bathroom renovation franchise firm that specialized in quickie renovations—acrylic overlay to existing bathrooms—bought a bunch of his crescent rods, enough to keep him going. And then, finally, in 2002, when he'd changed direction and approach so many times that he practically had whiplash, Starwood, owners of the Westin chain of hotels, called him. It had made a huge investment in mattresses after researching in minute detail what was necessary for a perfect night's rest, and the result was its trademarked Heavenly Bed. Now the hoteliers wanted to do something equally great with their bathrooms. They liked Moore's rod but wondered if he could test it for durability—maybe with a machine that would simulate the curtain being opened and closed 10,000 times?

"You're talking to the machine," Moore chuckles today. He personally opened and closed a curtain on a curved rod 10,000 times and could proudly report that it was up to the test. That first order from Starwood was for something like 20,000 rods, and he knew he was on his way because "nobody copies the Motel 6, but everybody copies the guys at the top of the food chain." Sure enough, pretty soon he had an order from Hyatt, and the curved shower rod was en route to becoming an industry standard.

Sales doubled every year, except for the year when they quadrupled. Which sounds great but actually meant he was constantly running to keep up with production. There was plenty of volume, yet every cent was going back into the business. And then, finally having built up momentum (and with that sincerest form of flattery, imitation, also appearing on the market), Moore sold the business in 2006. He won't say precisely what he was paid but does allow that it was in the upper teens of millions. He's not exactly Bill Gates, but he's definitely set for life.

"You know, my dad used to tell me, if you build a better mousetrap, the world will beat a path to your door. But that's unmitigated nonsense. You build it, then you have to beat a path to the world," he says. Perhaps the struggle, the personal sacrifices and the constant scramble to look for new ways every time he encountered a roadblock made him savour the victory even more. "It's still a thrill to go into a hotel and see one of my rods," he says by phone from Vietnam, where he is hard at work on his next venture, making plastic buckets. (And if that sounds dull, you haven't heard Moore talk about it; he's figured out a way to produce them faster and more cheaply than anyone else.) Vietnam is familiar territory to him because he flew helicopters in the Vietnam War. That year shaped his world view and prepared him to be an innovator, he says. "It changes your outlook to have combat experience. You don't view risk the same way as some others might. It doesn't matter how bad your finances are, so long as nobody is shooting at you."

CHANGING YOUR OWN RECIPE

For people who come up with innovative new products and businesses, personal reinvention is sometimes the impetus. They aren't necessarily looking to make the world a better place but looking to make their own lives better—more interesting, that is, not easier. Easier is sticking with the status quo. Getting to "more interesting" requires stretching past what's safe and predictable and venturing into the unknown, to learn something new.

That's Jean Blacklock's story. She'd already morphed once, from lawyer to financial planner/author, when, in her late forties, she decided to get into cupcakes, a career change that was greeted with . . . condescension. She didn't let it bother her—not, say, the way it would bother her if you said her red velvet cupcakes were dry. But standing in one of her sunny Prairie Girl Bakery shops in downtown Toronto, the scent of sugar and vanilla thick in the air, the trim and straight-talking Blacklock still seems bemused by the initial reaction. Unlike the skeptics, she understood that while a cupcake may be a fairly frivolous indulgence, figuring out how to make extraordinary ones, then market and sell them, would be a serious, life-changing gamble.

Her openness to new experiences—Blacklock is big on finding out what happens next, and seems to view life as something of a page-turner—helped motivate her to roll the dice. So did events in her personal life. A senior planner who liked her job at a big bank in Toronto, Blacklock fell in love with and married another executive, at which point it was suggested by the powers that be that one of the newlyweds should probably seek work elsewhere. She seized the moment, though she didn't really have a clear idea where to go next. One obvious option for a lawyer with financial expertise was mediation—a respectable choice that would keep her income about the same (and would, at cocktail parties, elicit zero condescension).

But sitting in a week-long mediation course at Harvard in the fall of 2009, she found herself daydreaming about cupcakes. Growing up on the Canadian Prairies, Blacklock had loved cooking and always imagined she'd do something in the food industry. Now, in class, she was doodling ideas for cupcake flavours in the margin of her notebook. She did some research and discovered that Toronto was one of the only big North American cities without a dedicated cupcake store downtown. This didn't make sense to her. Little iced cakes weren't exactly a trend and had been popular in other big cities for more than two decades. And then, a few weeks later, as she worked on her third estate planning book, the eye-opener: she caught herself rewarding seven hours of writing with an hour of thinking about opening a cupcake store. Specifically, the bags—she wanted clear plastic ones that featured the logo of her imaginary business. She did a little more research and discovered that bags like that were surprisingly difficult to source. The quest hooked her.

"I've always been a very driven person," she explains with a laugh. "When I was an articling student, I wanted to become an associate. When I was an associate, I wanted to become a partner. When I joined the bank and was running its Calgary unit, I wanted to run the Prairies. I've always been very next-stop oriented."

However, very early on she gave up on her real passion, baking. Her parents, entrepreneurs who bred Herefords and ran an auction market in Saskatoon, persuaded her that a law degree was a better idea. She could always go back to cooking one day, they told her. "But you so rarely do that, do you?" she says now. Once she had her law degree, Blacklock fell into estate planning quite by accident—mid-recession, the firm that had hired her out of university offered her a choice: take a job in the tax department or take the exit. Her father was blunt: "It's always better to pay the rent. Take the stupid job." Estate planning might sound dry, but for Blacklock it quickly became interesting because it was all about people's lives, desires and fears. "It's intimate," she points out. "You get what's going on with a family, with a couple, with someone with a terminal illness." And there was the satisfaction that comes from having helped people figure out something complicated and important. "Clients walk away thinking, 'Wow, I'm really relieved to have *that* done.'"

There, and later at Bank of Montreal's wealth management unit, she dreamed up new programs, like one to teach young people financial fluency, as an outlet for her creativity. "Wherever I am, I think, 'What can I do that's different?' That's always been fun for me," Blacklock explains. She rejects the standard narrative wherein a disgruntled Bay Streeter finally finds happiness by checking out of the corporate world. She was never disgruntled, never unhappy. She always found some way to make her jobs interesting for herself—she viewed that as a personal responsibility—and she loved working with people and managing teams.

Nevertheless, food had always been her dream. And so in 2010 she decided that "at this point in my life, I believe in myself enough that I can follow my heart." She knew it sounded hokey, but . . . making a living selling cupcakes really *was* her dream. Once she started trying to make it a reality, however, she quickly realized that some of her ideas were a little half-baked. "I had to make a decision. Did I want to do the baking and basically earn a small income and work hard all day, or did I want to create a business? And I decided on the latter." She might enjoy cooking several dozen cupcakes a day, but not dozens of dozens, day after day. "If you really want a store, you can't do all the baking. It's like having a ladies' clothing store and making all the dresses." When she really questioned herself, Blacklock recognized that her underlying motivation was less to frost cupcakes than "to build something that would reflect me, where I could be the type of leader that I really wanted to be," the kind who doled out praise and feedback right away, who was able to make her staff "feel proud to be putting cupcakes in a box, and make the customer's day a little happier."

Once she really committed to that vision and started hiring bakers, she knew she was "past the point of no return."

She was not, however, scared when she opened her first shop in the fall of 2011. "I'd done careful market research, I'd sweated every detail, I had done my very best at finding a good location," she says. She believed so completely in what she was doing that fear simply didn't come into it, just as it hadn't when she left her first marriage. Back then, she remembers, "So many of our friends and family were just shocked, and everything in my life changed. I look back and go 'Wow, that was courageous of me.' I should have been scared, but in the moment, you're just so sure."

Now on any given morning in her two Toronto shops (soon there will be a third), as many as eight bakers mix up dozens of fresh cupcakes. Watching her head baker ladle rich, creamy icing into a piping bag to ice a tray of chocolate cupcakes that have cooled, Blacklock notes that curiosity and flexibility are essential for running a small business. There's so much minutia, so many trivial yet important puzzles, she points out, such as how to keep the bakers' uniforms clean. Plan A: they washed the uniforms themselves; plan B: Blacklock washed them herself because she felt guilty asking staff to do it at the end of a long day; plan C: she found a laundry service and outsourced the hassle. "For the first three or four months," says the mother of two teenagers, "it was total exhaustion. More exhausting than when the kids were babies."

Eighteen-hour days were the norm, day after day after day, partly because, although she'd done her research, she'd underestimated the appetite for cupcakes and never anticipated success from day one. The original business plan called for Prairie Girl to sell fifteen dozen cupcakes a day.

It was something of a shock to discover that they could sell that many in an *hour*. On Prairie Girl Bakery's first Valentine's Day, it was a crazy scene, with customers lining up for more than half an hour to buy cupcakes—11,000 of them in total—as fast as the bakers could frost them. Blacklock wound up making buttons for her staff that read "I survived Valentine's 2012 at Prairie Girl."

Along the way, she's innovated in ways that seemed minor at the time but have turned out to be very important. For instance, she noticed that other bakeries often sent customers out the door holding a flimsy box, or with their baked goods concealed in an anonymous paper bag. Blacklock persisted in her quest to source the longed-for clear bag bearing a Prairie Girl logo and finally found a supplier who shared her dedication to beautiful and functional packaging. Carrying their cupcakes in elegant clear bags, her customers became walking advertisements for her products. "We hear this every day in the store: 'I've seen these bags around.'" Working closely with the supplier, she also created a box with a special insert to hold a single cupcake in place so its icing doesn't get smooshed. As it turns out, customers care a lot about such things; they like being able to buy a single cupcake that is packaged and presented like a work of art.

Blacklock has made other kinds of incremental improvements, too. She's rejigged recipes, changed all the photography on the company's website, created new flavour combinations and, when it emerged that her husband (and chief taster) is gluten intolerant, even came up with some gluten-free offerings. Her most successful innovation, however, has been simply to shrink her product. Prairie Girl's head baker was dubious, but Blacklock insisted, and now mini-cupcakes outsell full-sized cupcakes four to one. It was also the mini-cupcake that helped the stores through a recent mini-crisis. A local paper sent one of the regular-sized cupcakes to a lab and then published the findings on its fat and caloric content, which were heart-stopping for anyone on a diet and a PR nightmare for the owner of a business whose motto is "Live life one cupcake at a time." Blacklock rallied the troops at her stores with a pragmatic battle cry: "We aren't selling broccoli! Our aim is to sell the best cupcake, with natural ingredients, and on that score, we can be proud of what we do." And there was already a safety net in case sales flagged. Those mini-cupcakes—which, as even the critical newspaper article noted, were one way to satisfy a craving.

"To keep a business fresh," she observes, "there's constant innovation. It's not like, 'This formula works, so we'll leave it as it is.' It's more, 'What's the customer saying to us?'" To hear that for herself, she works front of the store at one of her locations every Friday for twelve hours. "That's where I get my ideas and see what's going on." That ongoing contact with customers is how she came to understand that they really want the unadulterated cupcake experience and aren't in the market for a healthy-ish muffin–coffee cake hybrid of the sort that Prairie Girl used to offer early in the morning, called the Early Bird. They were delicious, and Blacklock spent $500 on a lovely metal sign to advertise them and the staff were all specially trained to sell them, but a funny thing happened. Customers didn't want them, not even at 8 a.m. The few who did order them then gazed covetously at the rows of regular, decadent cupcakes. "Mine doesn't have frosting?" someone asked plaintively one morning after being handed an Early Bird. Blacklock conceded defeat, and that was the end of the Early Bird.

Despite the missteps and constant tiny details that require attention, figuring out how to keep the business expanding and improving isn't a chore. "I feel so alive! I just love what I'm doing," says Blacklock. "Absolutely the best part is building a team. I'm proud of the fact that there are thirty-six people in the world whose livelihood depends on Prairie Girl Bakery. That's an amazing feeling." But she's managed much larger teams—why is this so different? In a big corporation, Blacklock replies, the CEO leaves and there's a going-away party, yet afterward, the company carries on much as always. "But with your own business, you definitely know that people are looking to you for the direction and the strategy."

And she likes that but is already thinking that Prairie Girl is not her last act in the business world. In the future, five or six years from now maybe, she'll do something else. Philanthropy, possibly. "My only frustration with life is that it goes too fast and it's too short. What else could I do in my remaining ten, twenty, hopefully thirty years? What's next for me?" she says with the enthusiasm of someone who's just getting going.

Sometimes, she says, young lawyers who know her story come into the store, see her behind the counter and moan, "I'd love to leave the law." She hands them a cupcake with a smile, but she's thinking, "Why are you saying that? Either find something in law that you love or leave." People who aren't happy in their work aren't curious enough about trying to change it into something they could love, she thinks, or aren't curious enough about what else they could do. Personal reinvention is, she's discovered, not nearly as scary—and quite a bit sweeter—than most people think. And if you have to innovate not just in your own life but in your business in order to get there, so much the better.

CHAPTER EIGHT

Get Engaged

My first real job, not counting my newspaper route or babysitting, was at a Dairy Queen in Winnipeg. It was a tiny place, probably 500 square feet in total, about half of which was "behind the counter," including a walk-in cold storage room. My friend Emma worked there that summer as well, and when we were scheduled on the same shift we had a blast. We thought it was hysterical to respond to the little dinging bell from the front door—the signal that a customer had arrived—by racing each other out the *back* door into the alley; the loser had to serve the customer. I had no concept that my job performance affected anyone else or could have an impact on another person's ability to make a living (let alone my own ability to earn money). I fancied myself something of a rebel, a free spirit born to challenge authority. I was all of fourteen.

And then toward the end of August my work ethic hit a nadir. It was one of those perfectly still and hot late summer evenings when you are acutely aware your days of freedom are drawing to a close, and there was a party to go to. But I had to work, which meant being stuck at the restaurant until 9 p.m. I just couldn't bear the prospect. My solution, which at the time seemed genius, was to send my identical twin sister, Adrian, in to work my shift. We had actually never tried to swap identities before. We were too busy making sure people could tell us apart. But that night I decided it was the perfect time to try. Adrian was the alpha of our pack—for her, our friends would wait until nine o'clock to go to the party. This just wouldn't have happened for me.

So into Dairy Queen she went. Amazingly, she was able to fool the assistant manager for more than an hour, until a customer arrived looking for a soft ice cream cone. Now, there's nothing particularly difficult about making a soft ice cream cone. But somebody has to show you, at least once, how to use the soft-serve machine. Poor Adrian ruined three cones in a row before the assistant manager turned to look at her with new eyes. "Who *are* you?" he asked, and it wasn't part of a SWOT analysis. She was sent home (and we went to the party), and the next day I was fired. It was a story I told for years as a funny anecdote, though always with a little feeling in the pit of my stomach that I had been a jerk, especially to the manager and owner. They weren't just fooling around at Dairy Queen, like I was. Their livelihood depended on the franchise's success.

Like all disengaged employees, I just didn't care about the connection between my actions and the consequences for the business. So when issues arose at work, with the cash register, say, my response was "Oh well. Not my problem."

Unfortunately for my employer, I could afford to act this way—I was still fully supported by my parents, just working for extra pocket money. I was much less interested in doing a good job than in having fun and acting cool.

I was exactly the kind of employee who makes micromanagers feel justified in controlling all decision making. Give someone like teenaged me any say or input and I'd cause trouble. Or so the thinking goes. In reality, when employees are allowed to question authority, they are more likely to feel invested in their jobs and therefore more likely to perform them well. And when they are actively encouraged to question the status quo—when their viewpoint is sought out and valued—they are far more likely to think innovatively and come up with the sorts of ideas and feedback that help move an organization forward.

Now, I'm not trying to weasel out of taking responsibility for my bad behaviour. At fourteen, I certainly knew the difference between right and wrong. But it's also true that my lack of engagement was related to the way I was managed. When you're treated as though you couldn't possibly have anything of value to contribute, chances are good that you won't even try.

When people have a deeper sense of the purpose of their work, they tend to be more productive because they feel they're contributing in a positive way. You don't have to have a prestigious job to experience this sense of engagement, as I was reminded recently when, travelling on business and with a mountain of work to get through, I ordered dinner to my hotel room. The woman who brought it to me had a sparkle in her eye and a pleasant manner, and we chatted briefly as she arranged my cutlery and so forth. It was clear that it gave her pleasure to take care of others, and as she was leaving, she smiled and said, "I'm glad to have helped"—and she really had, because she made me feel cared for when I was far from home and overwhelmed by my workload. She makes her job meaningful not just for herself, but also for hotel guests, simply by connecting to it and feeling that what she does matters.

Many people, however, just aren't that interested in their jobs. As an adult, I've met and worked with plenty who have pretty much the same attitude I did as a teenager, who feel it's okay—justified, even—to slack off and cut corners on a regular basis. The important thing is not to get caught. Not only do disengaged people do their jobs poorly, they can also create a toxic atmosphere for everyone else. It's hard to feel enthusiastic about your work or to approach it as a challenge to do your best when you're surrounded by people who don't give a damn.

In a lot of workplaces, the majority of employees are pretty disengaged. Gary Hamel, a globally renowned management expert, notes in *The Future of Management* that research on employee engagement finds strikingly similar results in every country on earth. One study found that 14 percent of employees are highly engaged, 24 percent are disengaged (Hamel calls them "the maliciously compliant") and the rest—a whopping 62 percent!—simply show up for the paycheque.

That's a real issue if you're a manager trying desperately to innovate or if you're a keener in a "Who cares?" culture. Disengaged workers aren't invested in burning the midnight oil looking for the newest thing. They have no interest even in pitching in to help solve a company's minor problems. That's not how they take their own measure. And what's more, the truly disengaged worker not only isn't connecting to her job but may be subtly undermining the overall goals of the enterprise.

So what can employers do about it? Experts and employees alike agree on the answer: create an environment that invites or even demands employee engagement. Build it, and they will care. And the foundation is to create an atmosphere where people are encouraged to question the status quo and keep trying to come up with something better.

FINDING MEANING AT WORK

Michael Budman is the American-born founder of Roots, one of Canada's best-known companies. (The irony isn't lost on the laconic Michigander, who notes that, growing up, his favourite pastimes were hockey and paddling, making him a natural in his adopted homeland years later.) Canadian heritage is a big part of the company's branding and ethos, and being true to that heritage means, for Budman, manufacturing everything possible in Canada. That's not easy to do in this day and age, when labour costs in places like China and Vietnam are a fraction of what they are in Toronto, where Roots' leather goods are made.

But Budman is committed to keeping jobs in Canada. So committed that, a few years ago, when managers who'd been hired to oversee the business and introduce efficiencies recommended sending most of the work offshore, he called a meeting and "took a sledgehammer" to the whole team. Budman, whose bushy dark brown hair and wild eyebrows make him look vaguely Einsteinesque, has a twinkle in his brown eyes when he tells this story, the corner of his mouth lifting in a sly grin. Firing those managers meant he was right back in the full-time, hands-on role he'd been hoping to break away from, but, he says today, "It felt like the right thing to do. I had to re-engage."

And he and his partner Don Green had to be sure their employees were productive enough to justify higher labour costs. So they've spent a lot of time thinking about how to engage them, which is particularly tricky when it comes to piecework in the garment industry; sewing the same buttons on identical squares of cloth, day in and day out, just isn't inherently fascinating, even if you really love to sew.

The pair decided part of the answer was to create a pleasant, naturally lit workspace, so the Toronto leather factory is bright and airy, with huge windows. To encourage team spirit, employees are organized into "rinks" (yes, a hockey reference; each unit is divided by wooden waist-high walls reminiscent of an arena). In each rink, six to eight workers assemble an item—a purse, say, or a jacket. Each still performs a specialized task, so that one person always sews zippers while another inserts pockets. But they don't work on a conveyor-belt type system, where the item is passed down the line and individual employees never see the finished product. Instead, it's completed right in the rink, so that everyone who works on it gets to see the leather bag or jacket when it's all done and to feel a sense of connection to the end product. This, Budman and Green believe, helps ensure that employees understand the importance of their work.

And it does, says Maria Valente, a beautiful woman with clear, olive skin and curly brown hair she tames by pinning up. The rink system is why she likes her job. She feels connected not only to the items she sews but also to the other people in her rink. Shy but far from diffident, she chooses her words carefully when speaking English, a language she taught herself after emigrating from Portugal in 1995. Her brother already lived in Canada, and together, they looked for work in the garment industry. Maria's formal schooling ended at age thirteen when her parents pulled her out of school and sent her to apprentice, without pay, with a tailor in their small hometown. Maria doesn't describe this in tones of injury, but matter-of-factly. "I work because I need to work," she says, "and I'm so happy to have this job."

In recent years, Maria has taken over the production of samples, items that are used in promotions and in ads, or given to celebrities as gifts. She is proud, she says, when "her" garments are featured in magazine shoots, but adds, "I'm always trying to get better. There is no, 'Oh, today I'm so good, tomorrow I don't have to be better.'" Maria is smart, hard-working and extremely skilled at a job that would bore a lot of people to tears. What keeps her energized and interested is the environment in the factory, which encourages her to start every day the same way by asking, What can I do better than I did yesterday?

It wasn't an accident that Budman gave her responsibility for the samples. It's part of his strategy to keep her interested in her job. In a nice virtuous circle, she has input into the design process, so she helps make incremental improvements to samples, and that keeps her even more engaged. As the owners of Roots recognized, what really matters to most employees is knowing that they matter.

THE FLATTER WORLD

In the industrial age, a hierarchical management structure was the status quo. Mass mechanization probably required it, frankly—someone needed to be in charge of, and accountable for, allocating resources, including staff. But today, traditional ideas about management can be a huge impediment to innovation.

The reason? The status quo bias, and our tendency to do things as they've always been done. In traditional companies where people move into senior roles because they're able to work well within the existing framework, managers may not be at all open to fundamental change, or even, depending on their personalities, to incremental improvements. Managers are often the gatekeepers of innovation, and if they're highly risk-averse, really new ideas don't stand much of a chance. One of the worst side effects of a traditional, hierarchical approach is that employees are less engaged than they would be if they felt they had more power or control—or if they had managers who were actively trying to get them to think more innovatively, like the ones at Canadian Tire or Four Seasons.

Employees can be disengaged to some degree even if they're intrinsically motivated. When you're intrinsically motivated, bonuses, fancy titles and perks, though nice, are not what keep you going. What keeps you engaged is your own interest. Doing the job is in and of itself rewarding. No one else needs to convince you that your work is important. You already believe that it is. But if you're not also empowered to implement changes, a certain passivity and complacency can set in when a problem arises. That problem becomes the status quo—just how things are—and you may try to work around it rather than try to solve it.

There are probably few employees more intrinsically motivated than nurses, doctors and other front-line health care workers in hospitals. Sometimes you or I might have a hard time connecting the details of our day to our organization's overall purpose—sometimes we may lose sight of what that purpose even is—but if you're a nurse, there's really no question. You're in the business of saving lives, and the line between your activities and that goal is direct.

And yet it's also true that one of the most potentially deadly things inside a hospital is those very same nurses—or to be more precise, their hands. Unwashed hands spread infections that can kill patients whose immune systems have already been compromised. So handwashing or lack thereof is a critical issue in hospitals everywhere, and millions of dollars have been invested (and are still being invested) in encouraging health care workers to do more of it.

All health care workers know exactly what a hospital's organizational goal is—prevent disease, promote good health—and what the stakes are. And sure, the programs, the

Wash Your Hands reminders posted on the walls, the happy charts showing progress, they've all helped. Nevertheless, nurses (and doctors, and just about all front-line workers) don't wash their hands between patients 100 percent of the time. In most hospitals, they do so only about 70 percent of the time, despite the fact that it is such an easy way to accomplish something they're incredibly motivated to do anyway, which is to keep people alive and healthy. It's a real puzzle.

But one regional hospital in northern Ontario stumbled on an explanation that shocked administrators. After probing the matter and considering all kinds of reasons for noncompliance, they discovered that nurses weren't washing their hands enough because the hospital's soap was taking a real toll on their skin. They weren't even necessarily making a conscious choice to avoid unduly harsh cleanser; only after careful digging and questioning did the hospital discover that its crummy soap was the problem.

The administrators' next stop: a trip to the procurement department, buried in the basement, without a patient or front-line health care worker in sight. There, workers were also highly motivated, but they had a different goal, and that was to save money. They'd been told repeatedly to try to shave what they could off operating costs, so they'd laid in a large supply of cheapo soap—the same soap that was now, indirectly, threatening patients' health.

For the people who run the hospital, the experience was a sobering reminder of the importance of clearly communicating overall goals so that front-line workers and those behind the scenes pull in the same direction instead of working at cross purposes. The further away employees are from the central activity of an organization, the less able they may be to connect their day-to-day actions and decisions with the true mission of the place—and the more important, consequently, it may be to engage them.

Getting employees to make the connection between what they do and the company's success is vital, as Issy Sharp found in his hotels. To provide an extraordinary service experience for Four Seasons customers, he realized, even the dishwasher who never interacts with a real live hotel guest should be able to picture a guest drinking from the coffee cup she's cleaning. Otherwise she simply won't approach her job with enough attention to detail. Part of engaging workers is empowering them to make decisions on the spot that promote overall organizational goals. If the dishwasher notices the new detergent doesn't do a great job cleaning lipstick stains off the cup, you want her to feel it's her job to do something about that, rather than waiting until a guest complains. The more employees feel personally responsible for and capable of addressing problems, the more likely they are to think innovatively and deliver the kind of value every employer dreams of.

Nurses who are engaged *and* empowered don't just stop washing their hands. They take the next step and ask themselves, Why am I resisting handwashing? Am I too busy? Is the sink too far away? Do I not accept the premise? Or is there another reason? What can I do about this issue? They question themselves, and then do something constructive with the answers, instead of waiting for a hospital task force to ferret out the crucial information: Hey, that soap makes my hands red and dry.

To feel empowered, though, you first have to believe that managers care what you think.

FLOCKING THEORY (OR WHY YOUR BOSS SHOULD THINK MORE LIKE A BIRD)

Peter Engstrom is a former military man who was national director of security policy for the U.S. Air Force before taking his knowledge about innovation to the private sector, much as Rolf Smith did. In his consulting work, Engstrom discovered that too often a control-from-the-centre mindset works against the kind of collaborative innovation that really can help a business thrive. He'd been puzzling over how to help managers accept employees' capacity for innovation when, one day as he was driving to work, he glanced up at a large flock of starlings overhead. He was fascinated by the way the group of birds flew as one, swooping and whirling and shifting directions in a second or two, but always en masse. The birds didn't crash into each other or fall out of formation; they all seemed to know precisely when to bank and dive to stay with the flock.

When Engstrom researched how birds did this, he realized there was a very human application, and the flocking theory was born. Flying in a flock, the birds on the periphery make the decisions for all. Birds on the wing head toward food or water, and away from predators. Those on the outside are at the most peril—most vulnerable to a hawk, for instance—but also have the best vantage point. By signalling toward the middle, those outside birds can steer the flock. Their reward: they don't have to fly on the periphery for long. Birds constantly rotate from that position to the middle of the flock, so everyone gets a turn as lookout. But those closest to the action, the ones flying on the edge, are always the ones that make the calls.

In business, the reverse is usually true. Take a big consumer products company, where decisions about products, marketing and even display occur at head office, then are transmitted out to the field. The sales guy placing the product in a local store has little discretion in terms of where it actually ends up, despite the fact that no one is better positioned to evaluate how the product is faring relative to competitors' products and why. By consulting not with head office but with other salespeople in nearby areas, he could make informed decisions about what to do—and then report them up the chain. Flocking theory in action.

Already, this is what happens at Whole Foods, where employees in individual stores have access to a wide range of the company's data. They can see each other's salaries. They can see how individual products are selling. And, using a local store-based voting system, they can give input into decisions about product placement and even the pricing of products.

Traditional managers tend to think that giving employees this much autonomy is dangerous, but the beauty of the Whole Foods approach is that by empowering front-line employees, the company essentially mandates a culture of inquiry where every worker is in the business of trouble spotting and solution seeking. Workers aren't going to be told by head office where to place the organic peanut butter; they'll figure out how to place the product so it flies off the shelves. Of course, sometimes they'll make mistakes, but acceptance of missteps is a big part of creating a work environment that encourages curiosity and engagement, and therefore is a key ingredient of fostering innovation.

Smart managers also do this simply by putting themselves in employees' shoes, which, if nothing else, communicates respect to workers. On an airplane not too long ago, I met a manager who works at Comstock Canada, an independent contractor that plans and builds huge infrastructure projects for other companies. Brian Arthur is a strapping man, even after shedding more than 100 pounds in the past year. Baby-faced but with a linebacker's physique, he looks like the former NHL hockey player that he is (drafted as an enforcer by the Philadelphia Flyers, he admits he spent most of his time in the penalty box). But Arthur is a tough guy with a soft heart and the emotional intelligence to know how to use empathy to motivate employees.

For a company like Comstock, a job is won or lost on the promise of productivity—literally, how much and how fast workers can produce compared to those at a rival firm. Getting workers to perform at the highest level is therefore central to Arthur's job, and he does it by trying to help them understand how their work influences the outcome of a project. The more people recognize that each small thing they do may have huge relevance to the whole, the more inclined they will be to connect with that small thing and seek to improve it.

Arthur is convinced that barriers between management and staff demotivate the rank and file. He makes it a point to be on site as much as possible, which is time-consuming but means workers can communicate directly—especially important if they've encountered a problem—and in return, they feel that someone in a position of authority actually knows who they are and cares about what they need.

At a job site for a nuclear reactor, on a sweltering summer day, Arthur marched into the nicely chilled trailers reserved for on-site foremen and yanked out the plugs on every air-conditioning unit. "I didn't want the bosses to be sitting in comfort while the guys were sweating it out," Arthur explains, wryly noting that he took some heat for the move, but the goodwill of the guys slogging it out under the broiling sun was well worth it.

Feeling a sense of connection—of really being heard—turns out to be a critical component not just in terms of the volume of ideas people come up with but also in terms of the quality of those ideas. When Rolf Smith, the innovation expert behind Me, Inc., was a strategic planner in air force intelligence, he was asked to develop a strategy to promote innovation. "Of course, all the old colonels are saying, 'That's not something we can do! You're either creative or you're not.'" Smith, however, was a big believer in the old adage that the solution to an organization's problems can be found within its employees' heads. And he was convinced that everyone has the ability to think more innovatively, albeit in different ways and at different levels. So he developed an electronic "suggestion box" and invited everyone across the whole military to submit ideas.

A young airman at Kelly Air Force Base in Texas submitted a practical if rather mundane idea: install speed bumps outside the barracks. Smith recalls, "The guy said, 'I'm on the night shift, but I can't get any sleep during the day because people speed down the street in their cars.'" It was a reasonable if not particularly inspirational suggestion, and Smith arranged for the installation of speed bumps. The airman's second request was equally prosaic. His phone calls to his girlfriend from the barracks' telephone were the source of derision, with anyone and everyone listening in and heckling. So, "Let's get a phone booth outside the barracks." Once again, the innovation team implemented the idea. And then, a few months later, the guy sent in another idea. But this one wasn't mundane—it was so important that Smith still can't reveal what it was. All he will say is that "it was one of the biggest operational ideas we ever got," and it solved a problem the air force had been struggling with a long time.

The lesson for Smith was obvious. Show people their ideas will be heard, and they'll be encouraged to keep coming up with them and eventually will get to the truly revolutionary innovative stuff. A lot of people just don't—or perhaps won't—do the hard work of thinking deeply about solving problems until they know someone will listen.

ENGAGING YOURSELF

Take a step back for a moment and think about some of these same concepts in your own life. Maybe it makes perfect sense to you that being more engaged at work will make you happier and more productive (which may create a virtuous circle where you are recognized for your contribution and thus enjoy the job even more). But do you take that same idea of engagement home? Pick a subject, from your creativity in the kitchen to what you volunteer for in the community to how well you listen to your kids. More engagement doesn't just make us better at things we do. Others benefit, too, when we're not just dialing it in.

Engagement means caring about what happens and how it happens, and being persistent in the same way a young child is when mastering a new skill. To be engaged, you have to risk failure and rejection, and you have to be willing to work hard. Most of us would rather hold back a little and protect ourselves. We have to be coaxed out of our comfort zone, but sometimes there's no one to do that.

Which is why we need to learn to do that for ourselves. What if your job isn't particularly captivating *and* managers aren't courting employee engagement? Well, you can still reap many of the same rewards—starting with enjoying your work, trying to improve it or yourself in some way and taking pride in your contribution—if you actively focus on trying to engage yourself. Experts recommend tricks to help people do this, such as painting word pictures for motivation. For instance, instead of thinking of yourself as working the counter for a fast-food company that's committed to recycling, you tell yourself that you had a hand in keeping 2,652 plastic bottles out of the ocean last year. And a lot of social science indicates that simply setting goals can bump up your interest in tasks that aren't in and of themselves particularly interesting. You can race the clock; you can turn tasks into games you play with yourself—"Let's see if I can file this expense report in half the time it took me last month."

And there's something else you can do: exercise your curiosity, both to create and maintain engagement. A friend of mine has a son with attention deficit hyperactivity disorder (ADHD) who attends an alternative school designed for kids with such challenges but nevertheless has a tough time staying on track in the classroom. She tells him that when he feels his attention wandering, he should stick up his hand and ask a question (perhaps because the school itself is rather unorthodox, students' questions are not only welcomed but actively encouraged). Since he's started doing this, his interest in school has shot up, and so have his marks. The simple act of thinking up a question helps him refocus and stay interested. It's also helping him to think divergently, which for many people with ADHD is already a significant strength. Their brains are designed beautifully to borrow ideas from one field and apply them elsewhere—a hallmark of innovative thinking.

Asking questions is also, in case I haven't made this clear yet, the ticket to innovative problem solving. One way to feel more engaged in your work is simply to make yourself wrestle with thorny problems, even small ones. The intrinsic satisfaction that comes from resolving problematic issues cannot be overstated—just think of how good it feels at home when you figure out how to fix the drawer in the kitchen that always sticks or when you finally devise a homework regimen for the kids that actually gets results. At work, you get that satisfaction and something else, too, as problem solvers tend to be rewarded with more and more interesting projects.

CHAPTER NINE

Play Well (A Little More Roughly, That Is) with Others

There is a field—not the kind of field you are used to probably, more an open expanse of untended, dry dirt—filled with children, many of them barefoot. They are running and yelling, kicking a soccer ball at dusk, with the ramped-up energy only little kids possess at this time of day. Many of them are dressed in worn T-shirts and shorts that have seen better days, probably on other bodies in other countries, where, when you outgrow something, you give it to Goodwill. But tonight, these kids are not thinking they have less than others, that somewhere else children are accustomed to full bellies and easier lives. They are just in the moment, alive to the joy of having a *real* ball to kick around. In the past, they've had to fashion one out of plastic bags, or even scour the ground for a round stone. But this one is the kind real soccer players use, only better. This ball also generates power—three hours' worth for every thirty minutes it's played with—enough to light a lamp, or power a hot plate or a fan or a cellphone.

A marriage of fun and function, the Soccket is far more than just a cool gadget. It's a playful solution to a real problem, cooked up by people with no extraordinary skills, expertise or technical knowledge. The most extraordinary thing about them is that they aren't much older than the soccer players.

If you could turn the clock back to 2011, you'd see four young women—girls, really—huddled in a dorm room at Harvard, wracking their brains for ideas. They are undergrads working on a final project in an engineering course, and they're not exactly poised for success. Actually, they're at a significant disadvantage compared to many of their classmates because all four are liberal arts majors and pretty clueless when it comes to engineering.

Remembering that session, Julia Silverman, now twenty-three, laughs about how far away she and her classmates were from inventing the Soccket. "If for no other reason than not to fail this class, the four of us just locked ourselves in a room and thought critically, 'Okay, what do we have to work with?'"

Their weakness: zero technical expertise between them—a problem when the assignment is to create a multiplayer online game concerning some aspect of public health. Their collective strength: all four have different backgrounds and perspectives—one is Indian, another Nigerian, and Silverman and the fourth young woman are Americans who've been studying Africa—and all are passionately curious about the developing world. They are also hell-bent on tweaking the assignment to make it more ambitious. They want to take it further, to a destination that's a little more real, and they want to get there by focusing on actual problems in developing countries.

Another weakness, but it's about to become a strength: they've already weathered a failure. The prof didn't like their first effort, a mobile electronic health-record system for developing nations. It was "too remote," he said, adding, "Don't try to change behaviours. You'll do better if you leverage existing behaviour."

And so they decide to take a big step back and forget about the end goal and their marks for a bit. Focus instead on existing problems and behaviours. "We really honed in on that," Silverman remembers, "and decided, 'Let's take something that's already there, that people are already doing.' People are littering, so there's pollution. There are a lot of car accidents; road safety is a big issue." Pretty soon this brainstorming session becomes a free-for-all, complete with interruptions, half-formed ideas and some half-baked ones, too.

And this act of stepping back, pressing that mental reboot of control-alt-delete and starting afresh gets them to the problem that really engages them: one and a half billion people in the developing world don't have electricity. Forget online games—these people don't even have a functioning light bulb in their homes. Which leads the group to a discussion about the kerosene lamps people in the developing world use. They're smoky, so it's hard to see and breathe if you're sitting in a little shack somewhere and one is lit. Plus, they're dangerous. Kerosene, diesel and wood-burning alternatives kill 1.6 million people a year. On a daily basis, the fumes from a kerosene lamp are as harmful as smoking two packs of cigarettes. This particular public health problem, they realize as they talk, has some really interesting layers. Kids who live without electricity spend hours gathering fuel; they don't have a lot of chances to play. And after nightfall, without light, they can't do any homework, or read, or just draw for fun.

Talking about it is a bit of a downer, truthfully. Too bad there isn't some way to approach problems like this from a place of positivity rather than abject need. Too bad there isn't some way to tackle this problem that's *fun*—the assignment is to come up with a game, after all. Silverman doesn't recall which one of them first raised the idea of kinetic energy—creating power through movement—because the session had become a little chaotic at this point. But she distinctly recalls someone saying, "Don't they do that with bicycles? Pedal to generate power?" And before long they're talking about what else could be used to create a kinetic energy machine. What about a soccer ball? Kids everywhere play soccer. "No, a ball is too small," someone objects. They argue about that for a while, then decide to test the idea anyway. What have they got to lose?

Their proof of concept involved one of those clear hard plastic balls designed to let hamsters run around the house, but instead of a rodent, they threw in a shake-to-charge flashlight, and voila! A quick-and-dirty prototype. Next they consulted some experts, friends who were majoring in engineering to find out how to wire a real ball. "We weren't thinking of manufacturing something," Silverman says. "It was, 'How do we make something pretty for our final project?'"

Bad news: their friends in engineering told them the energy a moving ball could create would be negligible, too little to be of much value. However, after some discussion, the team decided this could be a strength, not a weakness. They weren't looking to power a whole city, after all. They just wanted to come up with something practical that a regular person, a kid even, could use. Thinking small might actually make their idea *more* useful and practical. "These engineering guys are used to dealing with gigawatts and megawatts," Silverman explains today, "but we were talking about small yet meaningful packets of energy, just enough to light a lamp or charge a phone—not even all the way. Partially charging a phone is great."

Their lack of expert knowledge turned out to be a saving grace. "If we'd had any expertise, we would have stopped. We would have been discouraged by the parameters of the box. But if you don't even know where the box *is*, it's easy to think outside it." And they had a strong emotional connection to the idea of a toy that could help some of the people they'd met—and studied—in the developing world. "Our personal experience and our stories," notes Silverman, "compensated for the knowledge gap, the lack of technical know-how."

Every setback they faced merely reinforced their certainty that they were heading in the right direction because, in just about every instance, the problems and challenges wound up forcing further improvements to the product. For instance, they soon learned that if you take the delicate machinery required to transform motion into energy, put it inside an object and then ask people to kick that object with all their might—well, you'll encounter some breakages. Okay. They figured out how to make the mechanism tougher and more durable. It wasn't easy, since they didn't know anything, really, about kinetic energy. But they learned. This process continued until they had a soccer ball that actually worked. Silverman and one of her teammates, Jessica Matthews, loved their creation so much—and believed so strongly that it would help people to help themselves—that, after they graduated, they decided to commercialize the Soccket.

Today, 6,000 Sockets are being kicked around by kids all over the world, via Silverman and Matthews's non-profit company, Uncharted Play. Each ball can store up to twenty-four hours of power and costs $60, about the same as a standard, high-end soccer ball; at the moment, the balls are purchased by donors and shipped to the developing world, but they will soon also retail in the United States, and for each ball that's sold, another will be sent overseas. Corporate donors have materialized; even Bill Clinton has sung the Soccket's praises.

But Silverman still gets the biggest kick out of travelling to places like Mexico and personally handing kids a ball. "Before they even know there's anything cool about this ball they grab it and run away and start playing with it," she says, laughing. "And you're like, 'Wait! You don't even know what's special about it yet!'" And when the kids come back ten minutes later, winded and smiling, and she shows them what *is* special about it? Their faces light up, "their eyes popping out of their heads."

What happens after that is perhaps most rewarding,

though. "Invariably, five minutes later they say, 'I have a *better* idea,' and they start talking about other ways to generate power. It's amazing how short the time span is from seeing something innovative to starting to innovate yourself."

Children, of course, have no mental block when it comes to thinking of themselves as innovators; they fully believe they're astronauts, after all, and movie stars and Olympic gold medalists. Encouraging this innovative and playful thinking is, Silverman believes, "really critical to innovation but also to the development of these kids and their personal growth." If it's not encouraged, kids wind up, as they grow older, being afraid of looking foolish—and that fear switches off innovative thinking more effectively than almost anything else.

THE TROUBLE WITH POLITENESS

Brainstorming worked for the Soccket team, and the staff on our show are always hopeful it will work for us when we gather for what, in my workplace, we call "blue sky" sessions. We corral everyone who works on *The Lang & O'Leary Exchange* in a boardroom, order in lunch and hunker down to talk about how to improve the show. They're classic brainstorming sessions. All ideas are welcome, and the intent isn't to come up with small incremental improvements but rather bold, big picture, revolutionary stuff. "If we were going to rip the show apart and start all over again, what would we do?" is often the starting point. And great ideas do come up, at every single session. I always leave feeling energized and incredibly grateful to work with such a creative team of people. But the fact that I can't recall a single one of those blue-sky ideas at this moment tells you pretty much everything you need to know about our sessions. When they end, we all go back to our desks, turn to the pressing issue of the show we have to get on air that very day, and the ideas we've brainstormed . . . well, they pretty much vanish right into that blue sky. At least we got free pizza.

Why don't our sessions result in actual innovations, like the Soccket for Silverman and her teammates? Is it something about us?

Well, yes. For one thing, we know each other a lot better than the Soccket team did; we've been working together for years, and research on brainstorming indicates that while established teams tend to be comfortable with each other, they're much less innovative than teams of people who don't know each other as well. So one problem with our brainstorming sessions is that they include all the usual suspects—there're no wild cards in the bunch to shake things up.

But to be fair to us, we're also pretty typical. Although group brainstorming is common practice in many lines of work—and thought to be so beneficial that it's a key instructional method in many classrooms, from elementary school onward—study after study has shown that it's usually not the best way to devise new ideas and settle on new directions.

Over the past few decades, researchers have devised a lot of ways to measure innovative and divergent thinking. Sometimes they ask subjects to come up with as many uses as possible for an everyday object like a brick. Other times they ask purely hypothetical questions, such as, "What if people didn't require sleep—how would the world change?" And they also toss out open-ended but less abstract challenges, such as, "Think of as many ways as you can to preserve the environment." Regardless of the type of test, almost all the time, people who brainstorm on their own come up with more ideas, and their ideas are also more creative and more original than the ones dreamed up by groups.

And yet, those who've taken part in relatively unproductive group sessions continue to believe, mistakenly, that they're more creative in a team than they are alone. This "illusion of group efficacy" persists, social scientists say, because of what they delicately refer to as "memory confusion," where most of us overestimate our own contribution to a group brainstorming session because we conveniently forget that some of our brilliant ideas were not ours at all but were dreamed up by colleagues. We unconsciously steal credit, in other words.

Why don't groups come up with more and better ideas? There are all kinds of logical reasons that brainstorming en masse might be less productive than simply sitting down with a piece of paper and trying to figure out things on your own. First, there's safety in numbers; knowing you don't have to carry the ball all by yourself can encourage intellectual laziness. Then there's the self-censorship factor: you don't mention an idea you're not entirely certain of because you don't want to risk sounding like an idiot in front of everyone else. However, another problem is that you're not supposed to censor or question others. Rather, you have to listen politely when *they* wheel out bad ideas (who hasn't sat through a meeting where people respectfully mull over a clear non-starter of an idea just because the person who came up with it has a big title—or is so nice that no one wants to rain on her parade?). Finally, there's the tendency in groups to promote harmony. When everyone is trying hard to agree, it feels uncomfortable to volunteer an idea that takes the conversation in a completely different direction or challenges the consensus that's been building.

As it turns out, though, the biggest obstacle to group success is even more basic: we take turns. Researchers have discovered that delays aren't good for innovative thinking because "they can disrupt the activation of images" that precede and create new ideas, or "they can interrupt the continuation of a train of thought." Apparently, our ideas don't develop the same way if instead of blurting them out we have to keep track of what everyone else is saying and wait patiently for a break in the conversation. The Soccket team didn't take turns; in fact, they interrupted each other nonstop. But disruption can be hard to replicate in more formal settings, like work or school.

Fortunately, there are ways around this. In studies where participants brainstorm via computer and can type in their ideas at any point while also viewing other participants' ideas on their screens, groups *do* outperform individuals both in terms of quantity and quality of ideas. Team members spur each other on and spark off each other. The same holds true with the old-fashioned, low-tech method of writing ideas on slips of paper, as Canadian Tire did with Project Darwin and Rolf Smith does in his Me, Inc. thinking expeditions, and posting them where everyone can see. Groups are more productive than individuals so long as they don't always have to take turns communicating *verbally*. (Researchers don't remark on it, but this seems to be another area where kids instinctively "get it"; interrupting and shouting out answers without putting up your hand or patiently waiting your turn may be bad etiquette, but it's also likely to result in more creative thinking.)

So brainstorming itself isn't unproductive, as the invention of the Soccket proves. The problem really is the polite and respectful sessions that are encouraged in schools and most workplaces, where people are encouraged to agree with each other. Often, there's an explicit instruction to refrain from criticizing each other's contributions, thereby ensuring that the kind of healthy friction that promotes innovative thinking won't occur.

There are some pretty easy fixes to make the process both more efficient and more likely to result in innovative thinking. First, the smaller the group, the more productive it will be, especially if it includes people with quite different areas of expertise. A blue-sky session with people just like you may be more fun and feel more comfortable, but cross-pollination is more likely to result in innovation. Second, start the process via computer or paper, if at all possible, and pay attention to other people's contributions. In experiments where participants receive a simple instruction to attend to others' suggestions, they come up with more ideas—and the ideas are more original—than participants who don't receive the instruction. Third, don't be afraid to argue. Contradictions and disagreement often force people to go beyond the surface, to ask better questions, to reframe their ideas in more innovative ways. Finally, the narrower your focus, the deeper and more creative your ideas will be. If you have to consider a big problem, you'll do better if you subdivide it into smaller pieces.

AVOID THE RUSH TO JUDGMENT

And that's just the beginning, according to innovation experts, who say that if you really want brainstorming to work, you have to introduce even more structure, not less. To someone like Claude Legrand, idea generation isn't some freewheeling, loosen-your-tie-and-let-your-imagination-run-wild exercise. Innovation requires constraints. Instead of "Think outside the box, people; let's dream up a better widget," it begins with creating a box with highly specific and fairly rigid parameters: "How can we make a widget that will appeal to tweens?"

Generating great ideas in a group, Legrand continues, requires a clear agenda and goals, as well as a different way of thinking about thinking: innovation starts with divergent, not convergent, thinking. Convergent thinking is all about analyzing options and solutions in a rigorous, analytical fashion to converge on the best one and figure out how to implement it. Divergent thinking starts with questions, as many as possible. "Why would tweens want a widget? What if the widget were free?" The goal is to generate more ideas by letting your mind wander within the preordained constraints.

Figuring out which ideas are the most practical or least

risky to implement—all that should come later. The problem with a lot of group brainstorming is that that is where we *start.* We skip the divergent and head straight for the convergent, dutifully looking for the one right answer we can all agree on, the same way we were taught to do in first grade. And most of us are trying to keep our heads in the game, not the clouds.

That's just as true in TV as it is anywhere else. The fact is, a daily show like the one I work on is a bit of a grind—we have an hour to fill, it needs to include that day's news, and we tape at 5 p.m. every day. So between 9 a.m. and 5 p.m., we are running pretty hard to pull things together. We're figuring out which stories are the most important, researching those, booking guests and putting together production elements such as graphs and video clips. There's very little time left over for experimentation. On a good day, incremental improvements are what we aim for. A snazzier graphic, say, or a really cleverly written introduction. So when we get this same team together to "blue sky"—well, it's pretty hard to avoid convergent thinking. We already spend all our time together, talking about the show. Pretty naturally we head straight to "How?" instead of pausing at "Why? Why do people watch our show? Why do we cover things in the way that we do? Why don't we do things differently?" We don't ask these questions because we think we already have the answers.

Another part of the problem is that, since we're taking this time out from our regular activities, we want to make sure it's worth it, and even though we set out to think really boldly about new ideas, it means we are also intent on accomplishing something concrete. So we rush to come up with plans that can be implemented right away, drifting away from the original and back to the mundane. But that's not the right strategy if the goal is innovation.

You have to free your brain to roam to places that are a little impractical, and innovation consultants have come up with some great ways to encourage that. One of my favourites comes from Legrand, who tells people in group brainstorming sessions to try to come up with the *worst* possible ideas they can think of. For instance, if the question is "How do we attract seniors to our bank?" some of the worst ideas would be "locate ATMs in dark alleyways with thugs lurking nearby" or "reduce the print on the keyboard so it's barely visible." Once you have a list of really, really bad suggestions—and coming up with them does force your brain to work in a different way—you try to flip them over into the positive: "Locate ATMs in well-lit areas with security people standing guard."

There are many other ways to create this fresh-eyed perspective in our everyday lives. When I was at university, one technique the instructors in the architecture program used was to tell us to draw an object—a candlestick, for instance—and then they'd turn it upside down and have us draw it again. Without fail, the upside-down version was miles more accurate. The reason is simply because our brains stopped classifying the object as "candlestick" and simply perceived a bunch of lines and curves, which we were then able to copy more accurately. Think of all the times we mentally file an idea—or even an impression of a person—into an invisible filing cabinet, then throw away the key. If we could turn the idea (probably best not to try it on a person) upside down, what might we see differently?

Of course, all these exercises take time, and generally, especially in work situations, there's pressure to produce as quickly as possible. But there are two compelling reasons to give divergence time to work its magic. First, researchers report, "original ideas will usually only be generated after people have suggested the more obvious ideas." If you try to short-circuit the process, you may never break through to discover the original and innovative ideas—which might include reframing the entire question and attacking from a different angle altogether. Second, studies show that "people tend to select conventional ideas [ideas that are feasible but not original] out of the pool of ideas generated during a brainstorming session." We drift toward what feels familiar and doable—closer to the status quo—without really exploring the more "out there" possibilities.

PLAYING TOUGH ISN'T PLAYING MEAN

Disagreeing with and challenging each other—as the SawStop guys do, and the Soccket inventors did—is another aspect of just about any successful collaborative innovation. If you want to be sure you're headed in the right direction, there's no better way than having to defend and debate it. This is not, however, the way we're taught to brainstorm at school. "Listen respectfully without interrupting" is almost always the rule.

Somehow, as a society, we seem to have decided that the

heat of battle is toxic for children and that losing is so debilitating that we should pretend it doesn't happen. Hence, the trend across North America toward non-scored sporting events where, kids are told, the emphasis is on fun, not winning. Of course, they keep score anyway. At a "fun" soccer game, there will always be at least a few children protesting at the top of their lungs, "That's not fair! Our team *won*!" Shushing them and insisting that there are no winners or losers is a terrible idea, in my opinion.

Not only is striving to win a worthy goal—what problem could there possibly be with exhorting kids to do their best?—learning to lose is also incredibly important. Learning to lose a game (or an argument in a brainstorming session) and get right back up and play again with heart doesn't just build character. It builds the type of skills necessary for believing in your own capacity—both to take risks and to survive and learn from mistakes. The skills, in other words, that innovators like Addison Lawrence, Jean Blacklock and the Soccket team have in abundance.

Losing and being gracious about it (as well as being quietly determined to try harder and do better next time) doesn't come naturally to kids. That's precisely why we need to help them learn to do it in brainstorming sessions in social studies classes, or on the softball field, where the only thing that's really at risk is their pride. Learning to lose there, where it's safe, means they won't be crushed or bewildered by losses later on, when the stakes are a lot higher. Kids need to be trained to dust themselves off, figure out what went wrong and try harder next time, and nothing teaches those lessons quite so well as the occasional defeat.

Protecting them from the consequences of competition won't make them more competitive, though it could turn them into sore losers. Remember the American CEOs who instantly perceived something different about their Canadian employees—a fear of conflict and of failure, no matter how small? The lessons of the basketball court are hard to leave behind.

Striving to do better than the person next to you—*truly* better in a fair contest where competition, not an exercise in self-esteem promotion, is the point—is a powerful motivator. Without points of comparison and chances to measure ourselves against others, it's hard to be self-aware. You could do a SWOT analysis and blithely conclude you had no weaknesses and faced no threats. This does happen, as a friend of mine who works in wealth management discovered recently when she evaluated her young assistant, using the same standard form he used to evaluate himself. He was shocked when he compared the evaluations. He'd rated himself as exceeding expectations, across the board, whereas his boss had rated his performance as merely average and in some areas needing improvement. He simply didn't know how to cope with this feedback, she told me. It was as though he had never been told that anything he'd done was less than perfect, let alone average or even subpar. He had no idea how to pick himself up and focus on using the "loss" as a spur to become better.

One of the most powerful lessons of the innovators I've met, whether young and idealistic or old and methodical, is that they've enjoyed the journey not *despite* its difficulties, but almost *because of* them. They relish not only the war stories but also the actual one-step-forward-two-steps-back dance of innovation. To them, it's fun, and that's why they don't stop with one innovation but keep pressing forward, looking for other ways to innovate. Challenges, detractors, missteps, losses? All part of the rough-and-tumble of the game. To hear them tell it, they're playing, in a way, all day long. And although they don't always win—because, just as we tell our kids, you can't always win—they are always engaged and fully alive and curious about what will happen next.

CHAPTER TEN

Talk to Strangers

It's the mid-1990s and three men are riding in a car. Colleagues, they have known one another less than a year; one is Spanish, the other two, American, and they've been thrown together by their employer, Otis Elevator, to puzzle over a problem. The goal: Find a way to hoist elevators using smaller and lighter technology.

These men met for the first time not long ago, but since then they have spent hours a day together in the same lab space, where they've tacked funny and inspirational pictures on the walls. They have learned each other's kids' names and are now familiar with one another's subtle nuances of humour. They're not making all that much progress on the design project, but they're having a great time working together, not least because each brings a different perspective to the table, which has renewed their sense of engagement with their work. It seems stimulating and exciting again.

But today they're taking a bit of a break, driving to a test site to check out experiments on a new kind of cable that Pedro Baranda, the Spaniard, is excited about. Developed in the chemical industry, it's fibre based, not steel, so it can be fabricated into a much thinner cable while retaining the same tensile strength—a necessity when you're hauling heavy elevators up and down. In the car, there's music playing, but suddenly no one's paying much attention to the song because one of them has just had an idea that could change everything, an idea so simple that it's electrifying. For the rest of the ride there's only one topic of conversation: "Will it work?"

Otis is the world's most successful elevator company; every three days or so, the equivalent of the world's entire population rides on its elevators and escalators. Innovation has been part of its corporate culture since Elisha Otis founded the company in Yonkers, New York, in 1852. He'd come up with a pretty big incremental improvement: the first "safety" elevator, so passengers wouldn't plunge to their death if the mechanism—at that time, a rope over a pulley—failed. From its inception, Otis Elevator set the standard for safety, and later for ride quality (the less noticeable the movement of the elevator, the higher the ride quality).

But elevators are the nemesis of many architects and engineers. Their workings take up valuable space inside a building, and it's also very hard to make the top of a building look elegant if an elevator is part of the plan. The machine room that houses the giant gears and wheels must by necessity be taller than the building, and so it protrudes in an unsightly way. Various solutions to the design problem have been offered over time, but they all start with the same assumption, that an elevator has to be hoisted on cables, using a gear system.

In the mid-1990s, Otis—now part of a global conglomerate, UTC—decided to bring all of its researchers worldwide to Farmington, Connecticut, where they could work together in R & D labs. There were Russians, Germans and Italians, there were Japanese and Brazilians. And there was Pedro Baranda, who has a Ph.D. in material sciences and was familiar with the United States, having studied at Rutgers. (His wife, however, needed a little persuading to leave their home in beautiful San Sebastian in northern Spain. They arrived in Connecticut in July, so the harsh reality of winter in the Northeast didn't make itself known for several months, but when it did, she made it pretty clear that it was not to her taste.)

Nevertheless, for Baranda, it was an exhilarating time, because working with such an eclectic group was changing his perspective on elevators. The Americans, for instance, focused on high-rise elevators that could quickly ascend dozens of storeys; the Europeans were more interested in building low-rise elevators, thus reducing the size of the shaft and housings. In Japan, energy consumption was a focus, as was ride quality. Being exposed to different points of view was like pressing fast-forward mentally—not just for Baranda but for everyone working in R & D.

Baranda's own particular interest was the new fibre that chemical companies were developing for cables. The fibre was promising because of its D:D ratio: the diameter of the cable was small, and therefore the diameter of the pulley could be much smaller than usual, too. This meant that gears could likewise shrink. It seemed like the route to devising an elevator that not only had more compact workings, but also was more energy efficient. Baranda and the two American researchers he'd been grouped with began experimenting with the fibre, sending it to off-site labs for field tests, as well as taking it home and instructing their children to play rough with it—to try to destroy it, even.

But then came one problem after another. The fibre was tested in a fairly standard fire scenario and it burned in a way steel cable never would. Next, as Baranda explains dryly, "Someone picked up some scissors." The fibre is heavy, so not exactly easy to cut through, but it can be done. The trio reluctantly turned back to steel. But as is so often the case, these setbacks and "mistakes" became important turning points. They were not, as they first appeared, dead ends.

The group had grown accustomed to the way the fibre moved—it was beautifully light and flexible, like a rope. Once the trio moved back to steel, they had to abandon the idea of a light cable. There's no getting around the fact that steel is heavy. But the concept of flexibility had lodged in their minds so securely that on that car ride, suddenly they were talking about a flat cable. Baranda can't recall which one of the three first said the words, and he doesn't think it's important. It was a group effort, intellectually, with each person feeding off and adding to the ideas of the last. For weeks thereafter the trio talked about the "flat rope" until finally one day one of them realized, "That's a *belt* we're talking about." And sure enough, it was. The beauty of a coated steel belt is that it can be as wide as necessary for strength, yet remain very thin (think of the belt you wear with pants).

Just like that, the small team had solved a major design problem, because once you're using a flexible belt instead of rigid round cables, you no longer need huge gears—or an unsightly machine room perched on top of a building. "I have pictures that hang in our boardroom of me with my arms spread far apart to show the width of a normal pulley system, and then my finger and thumb touching, to show the width of our new system," says Baranda, who is now president of Otis Worldwide. "It looked so radical that we just patented it and started doing some lightweight experimentation. We realized we could make the smallest machine in history if this technology could work."

And it did. Several working prototypes later, the Gen2 elevator was born. It's the fastest-growing product in the company's history and also the most innovative. The Gen2 doesn't require a machine room, so it increases the available space within a building, and the belt that moves it actually *creates* energy. The elevator still needs external power to ascend, but because a belt can rely on gravity to help with the descent, the Gen2 generates power as it goes down, in much the same way a turbine does. Another plus: ride quality is also better—it's smoother and quieter.

The key to the breakthrough was putting together people who had very different views about what an elevator is and what it should be. Their ideas and clashing opinions bumped up against each other in ways that nudged each group member out of his own comfortable mental groove and forced them all to rethink the basics: Why *did* an elevator need a pulley, anyway? Just as Canadian Tire workers wiped their mental slates clean on the subject of men, the team at Otis pressed control-alt-delete on the topic of elevator engineering. And they couldn't have done that without each other.

ON THE OUTSIDE, LOOKING IN

Being surrounded by different perspectives and immersed in different experiences truly is mind-expanding. For instance, living in a foreign country is strongly correlated with thinking innovatively, researchers report. People who've studied and worked abroad are more creative idea generators and problem solvers; they do better on tests of ideation and on tests that require lateral thinking and making unusual associations. Simply travelling in foreign countries doesn't affect creativity, though, which makes intuitive sense. If you've been immersed in a foreign culture rather than hanging out in the lobby of the Hilton waiting for your air-conditioned tour bus to arrive, you've been exposed to different foods, different accents if not different languages, different values, different geography—a different way of life.

All of that is bound to have an impact on your mini-c if not your small-c creativity. People who've had a stranger-in-a-strange-land experience are not only more creative, they're also almost certainly more curious—they've had to be, simply to figure out what to bring to a dinner party or what not to say in a job interview in a foreign country. Curiosity is likely what got them on the plane in the first place.

Throw together several outsiders—like those on the Otis team or the Soccket team—and you secure a real innovation advantage (so long as everyone can communicate, that is). Why? Because people who've lived abroad don't just do better on the type of pen-and-paper tests of creativity that researchers use. They also lead the pack of innovators in the real world, according to *The Innovator's DNA,* the book Clayton Christensen co-authored with researchers who conducted a six-year study of particularly innovative individuals and companies. "Our research revealed that the more countries a person has lived in, the more likely he or she is to leverage that experience to deliver innovative products, processes, or businesses," the authors reported. "In fact, if managers try out even one international assignment before becoming CEO, their companies deliver stronger financial results than companies run by CEOs without such experience—roughly 7 percent higher market performance on average."

They've had the experience of being outsiders, and as we've seen, outsiders have a real advantage when it comes to innovation. Lou Gerstner at IBM, Stephen Wetmore at Canadian Tire, Steve Gass at SawStop, Jean Blacklock at Prairie Girl, Sean Moore with his shower curtain rod, the Soccket team with their power-generating ball—all were outsiders to the fields in which they wound up introducing innovations. Many had no formal training in those fields, and that probably helped them think more innovatively. Expertise might actually have blinded them to the problem they wound up solving, or prevented them from spotting the opportunity in it.

Outsiders approach a situation with fresh eyes and fewer assumptions. They may be entirely unaware what the status quo in an industry is, and thus are less likely to have a bias toward it or an interest in preserving it. To come up with new ideas, they don't have to wipe the mental slate clean—it already is pretty clean. And since they have a different area of expertise altogether, they may be able to "borrow" a solution or innovation from the field in which they have direct experience.

The world really does look different from 30,000 feet. Distance provides objectivity. But that's not all. Distance also improves creativity. According to what social psychologists call the construal level theory, anything that we don't experience as happening right now, right here, to ourselves is psychologically distant. And the more psychologically distant something is, the more likely we are to be able to think about it in abstract terms rather than concrete ones. Our minds are able to float more freely, past the contextual clutter, and make new associations—just as a three-year-old does when she picks up a banana, pretends it's a telephone and begins an animated conversation with her imaginary friend on the other end.

According to a fascinating and now extensive body of research, scientists have shown that it's possible to induce a state of psychological distance just by attempting to take an adversary's perspective, or by thinking of a problem or question as being pie-in-the-sky unreal, or by imagining that we're solving it for a stranger or far in the future. And the further away the problem or puzzle seems, the more creative people tend to be.

Consider a classic insight problem: "A prisoner was attempting to escape from a tower. He found a rope in his cell that was half as long enough to permit him to reach the ground safely. He divided the rope in half, tied the two parts together and escaped. How could he have done this?" People are most likely to come up with the solution—the prisoner unbraided the rope lengthwise, then tied the two strands together—if they are told that the tower is located far away, or that the prisoner was a stranger, or even that the problem was devised by someone in a foreign country. When people are told to imagine that they're the prisoners in the tower, the solution rate plummets; ditto if they're informed that someone two miles down the road came up with the problem. The closer to home a problem feels, the more likely we are to think of it in concrete, granular terms and miss the aha! moment of illumination.

Construal level theory doesn't simply apply to brainteasers. It also holds true when people are asked to generate ideas about modes of transportation, or to draw an alien, or to come up with a suitable birthday gift—when they're told they're doing this for someone who lives far away, the ideas come much more easily and are also more original and innovative than when they're told they're doing this for themselves or someone nearby.

This helps explain why outsiders have such a big advantage when it comes to innovation. They have more psychological distance because they're not all wrapped up in the nitty-gritty details of a problem. The research suggests that it's possible to get that outsider advantage yourself simply by thinking about a problem as though it's occurring in the future, or to someone else, or far away. And if that doesn't do the trick, try to surround yourself with people who are dissimilar to you or are actually strangers, so that the situation itself introduces feelings of psychological distance. As Pedro Baranda's team at Otis discovered, a little distance can go a long way toward solving "impossible" problems.

UNDERSTAND HOW STRANGERS SEE YOU

It may seem like a given that a company in the business of creating products for consumers would keep those consumers squarely in mind. But the battlefield is littered with fallen businesses, big and small, that forgot to do just that. Barry Jaruzelski, a New Jersey–based management consultant at Booz & Company, has spent thousands of hours studying many kinds of companies and has found that some of them simply don't think to ask customers what they want. They never venture outside their insular organization to find out what outsiders are thinking, or why.

He tells the story of a Silicon Valley computer company that came very close to blowing it for that reason, even though their high-performance servers were, by any objective measure, the best on the planet. Their servers were the fastest and had the largest storage capacity, yet the company was losing market share to rivals with inferior products. Jaruzelski was hired to find out why.

Is the problem that customers are stupid? he asked himself, then dismissed his own question. No, usually customers aren't stupid. Jaruzelski needed more information, so he asked the sales team what it knew about the customers. Very little, it turned out. The salespeople had never even visited a customer, even though one of their biggest ones, an Internet service provider, was just ten minutes down the road.

So down the road Jaruzelski went, for a little field trip with the sales team. Now, what ISPs do is run servers, thousands upon thousands of them in this case, routing traffic over their networks. For an ISP, a competitive advantage depends on greater speed and storage, and by those measures, Jaruzelski's client should have had a lock on the market. But the client's corporate focus wasn't the customer. It was the product: "They had this idea that if they built the very best servers they could, people would buy them." At the ISP, Jaruzelski talked not to the CEO but to the guys who actually use servers—the techs whose job it is to keep them running twenty-four hours a day in giant warehouse-like spaces filled with racks and racks of servers, and colossal air conditioners blowing in cool air so the servers don't overheat. When a server requires repair, as happens not infrequently, it's the techs' job to replace them on the fly, with no interruption to the service they're providing. In IT parlance, it's called "hot-swapping."

The techs told Jaruzelski and the sales team, "Yeah, your boxes run a little faster, but what matters more to us is how fast we can 'hot-swap' a server—yank it out of the rack and slam a new one in as quickly as possible to minimize the lost horsepower. And the thing we really don't like about your server is that the plugs are on the back."

Plugs on the back?! The server-maker had never given much thought to the placement of the plug. It was incidental, external—what *really* mattered, they "knew," were the inner workings of the things. Well, not to their customers. The placement of the plug is critical in a hot-swap. If the plug is on the back of a server, the tech has to race around the entire rack twice to make the change; if the plug is on the front, the tech can perform the operation standing in one spot and there's less risk of an interruption to service. Jaruzelski's client's narrow focus on improving its product's speed and storage capacity was hurting its business because, as he points out, "Innovation is in the eye of the beholder. It's what is valued by the customer."

Companies that assume they know exactly what customers want can wind up in serious trouble. This is what happened at Research In Motion. RIM's big advantage in its market was always security; RIM runs its own network, whereas everyone else has to rely on a phone company's network, which is de facto less secure. So when Apple started churning out hits like the iPhone and iPad, RIM banked on the idea that security-conscious big corporate customers (and governments) would stick with the BlackBerry, thereby giving the company enough time to nail a good consumer product that could compete with the iPhone. And RIM actually did manage to make small inroads into the consumer market because kids like the BBM—the instant messaging app—of the BlackBerry. But it looks like RIM may not have asked the right questions: Why do people like Apple's products? What do they want that Apple provides and we don't?

As it turns out, a lot of us care deeply about interface and design, and value it much more highly than security (and not surprisingly, Apple has been working hard to improve its products' security and may soon roll out some kind of software that encrypts data). But because RIM was so focused on appealing to the broader consumer market, it took its corporate customers for granted, simply assuming that what had worked so far—better security and a more user-friendly email system—would always be enough to keep them loyal. RIM didn't ask corporate customers what they needed because the company already "knew"—except it didn't. Corporate customers also want pleasing, streamlined design, so much that some major corporations and governments already allow employees to use Apple products at work. The iPad was the big game changer, in my opinion. I know one CEO who got an iPad and, the very next day, went into work and instructed his IT department to get them for all senior executives and to find a way to make them secure.

Of course, sometimes customers cannot articulate what it is they want. It's tacit-level information, meaning they can't necessarily verbalize it but they sure do know it when they see it. This is why so many consumer product firms resort to ethnographic approaches, observing people closely to try to figure out what really matters to them and how they use products. When Rotman business school dean Roger Martin was working with Procter & Gamble in his former life as a consultant, he decided to take ethnography to a new level to try to help the company's hair care division become more innovative. He carted the product team off to a salon to observe customers, and he also gave those customers cameras and asked them to record their experience at the salon. When the photos came back, there weren't a whole lot of still-life shots of shampoo and conditioner bottles. A great many of the pictures were of male hairdressers. Apparently for women, on a tacit level, getting their hair done is a sensual experience (even if the hairdresser happens to be gay). It's all about feeling pampered and cared for. This pointed the team in the right direction for innovation: it needed to focus on packaging and marketing that stressed luxury and indulgence. A lot of hair product companies have since copied the approach, but for a while P&G's packaging and messaging were unique in this way. And sales soared.

Sometimes, people don't know what they need until it's available to them. This is very often the case with service innovations, which usually just provide a new and better way to do something we're already doing. "The causal mechanism behind a successful innovation is that you've figured out that there's a job out there people need to do," says Clayton Christensen, the Harvard prof who's probably the world's leading authority on innovation. The jobs we have to do remain fairly stable over time—it's only how we do them that changes, he explains. "'I've got to get this package from here to there with perfect certainty as fast as possible.' Julius Caesar had that job to do, so he hired a horseman with a chariot. Now we have FedEx. The solutions change, but the job is still the same."

TRY TO SEE THE WORLD THROUGH YOUR CUSTOMERS' EYES (EVEN IF YOU DON'T HAVE ANY CUSTOMERS)

Virtually every process innovation starts with the same question: What do customers need and how can we produce it more effectively and cheaply? The most famous approach to answering these questions is Toyota's lean manufacturing methodology, which focuses on eliminating waste. At the risk of oversimplifying, here's the thinking: Let's say there are twenty steps to making a car—would the customer pay for each step? And if not, could that step be eliminated, or performed more simply and cheaply?

Lean manufacturing is one of the business concepts with the clearest applications to everyday life because it's all about efficiency. If you think through all the steps of making a family dinner, for example, there are probably a few you can eliminate right off the bat, since your "customers" simply don't value them. If your kids are anything like the ones in my house, for instance, I guarantee you they place no value on your cheese-grating skills—they won't even notice if you switch over to the prepackaged stuff and sprinkle that on their pasta instead.

But lean manufacturing isn't just about making your own life easier. To figure out what's important to your customers, whether they're your kids or actual clients, you need empathy. And in and of itself, some awareness of another person's perspective is almost certain to improve the relationship.

Lean manufacturing your life goes something like this: Step one, identify the product you're selling. Step two, identify your customer. Step three, examine the whole chain of "production"—what you do to create the end product. Then ask yourself whether your "customer" cares about every step of the chain.

It's well worth the time, says Christensen. He explained to me how he applies these concepts when thinking about his relationship with his wife: he thinks about "the job" that his wife has "hired" a husband for. Using a lean approach, he doesn't ask what *he* values or cares about, but what she, the "customer," values.

"What *I* think she needs from a husband is very different from what *she* thinks she needs," he explains. The lean goal, then, is to understand what she values and eliminate waste—wasted effort, for instance, performing what he views as key husbandly tasks but which his wife might not care about at all. She might not, for instance, give a hoot about getting a fancy card and flowers on Valentine's Day but might highly value having him do the laundry on a regular basis. Understanding his wife's criteria for satisfactory performance of his job has a few distinct benefits, he notes with a laugh: customers tend to "develop tremendous loyalty around 'products' that get the job done well."

Lean manufacturing your life can, then, become a form of innovation: efficiencies that lead to incremental improvements in your relationships with others. I've seen this in my own home. I'm a compulsive tidier, possibly because I lived alone until I was thirty-two and got used to doing things the right way—my way! Or perhaps I'm a bit of a neat freak because I grew up in a household with seven kids, two busy parents and a few pets, and if you left your stuff lying around, forget it, you'd never find it again. So I learned to put things where they belong, and today, I pride myself on keeping our house nice and tidy. To me, this seemed like something pretty wonderful that I do for my family—selflessly, because they aren't quite as grateful as I think they ought to be.

But there's another way to look at this. When I started thinking about tidiness as a service I provide, well, it was undeniable that my "customers" didn't value it. My product was a well-ordered home, and the chain of production involved steps like reminding the kids not to put their dirty feet on the white duvet covers at the cottage. (I know, I know, but trust me, white beds really look great in our cottage.) However, the only person buying any of this was me. Hyperneatness was *my* value. My "customers" had quite different needs, ones I'd never really thought about until I looked at the world through their eyes. The kids just wanted to be able to relax and sprawl out. For them, that's what it means to feel at home. A certain amount of order and cleanliness makes sense, but they don't want to live in a museum run by a drill sergeant with a feather duster in her hand. And I was doing something that made them feel less comfortable, less at home. So I learned to relax my standards. A little.

Thinking about what other people need from you as a job you are trying to perform can help you figure out the little changes that will make a big difference. And that's really all that innovative thinkers like Gordon Eberts or Issy Sharp did. They looked at the jobs out there and figured out new ways to get them done in a more streamlined and convenient fashion. And knowing what your customers need starts with putting yourself in their shoes, which requires empathy.

My on-air performance coach, a soft-spoken Texan named Nick Dalley, has been drumming this into me for about ten years now. Gently, because that's his style. He radiates kindness, actually, which is important in his line of work, since his clients can be pretty thin-skinned about how they come across on television. And insecure: a lot of Nick's coaching involves reviewing videos of yourself, which never stops making you squirm no matter how long you've been in TV, and then walking you through what you could do better. This process could feel like torture, but Nick's secret, at least with me, is to make us co-conspirators. We are a team, serving the same customer—the viewer. Everything we do together must focus on giving viewers what they need. Nick quite deliberately uses that word—*need*—because his theory is that people who watch the news need to feel taken care of by the people delivering the information. Viewers need to believe that their interests—getting information in a way that makes sense to them—are being served.

Thinking about my job this way has really changed how I approach it. I'm much more likely to speak directly to viewers now, for one thing, and for another, I try to follow Nick's advice and think not about myself and how I'm coming across but instead think about the audience and what they need. Here's the balancing act: never lose sight of the fact that you are the servant and viewers are the masters, yet always remain in control and convey competence. I think of myself as being sort of like a high-end gentleman's valet in days gone by; not in charge, but capable of providing good guidance. Think of the truly great on-air anchors and interviewers and chances are that, on some level, you think of them as friends. People you know and trust. I guarantee that those luminaries, from Ted Koppel to Peter Mansbridge to Diane Sawyer, aren't thinking about themselves when they deliver information or conduct interviews. They are thinking about you.

Which is as it should be, not least because it keeps them on their toes, always trying to find new and better ways to do their jobs. "What do others need from me?" is a question that, in the business world, drives successful innovation.

DON'T ASSUME OTHERS THINK THE WAY YOU DO

Many of us don't even really know how we think or the ways in which our thinking differs from others. We may compare ourselves to other people in terms of intelligence or aptitude or skill, without even recognizing that we have entirely different ways of thinking. Style of thinking may, in fact, be one of the last bastions of prejudice. We accept and even encourage diversity in so many other ways, but not this one.

Certainly throughout school there's not a lot of emphasis on letting people process information in the way that feels most comfortable to them. The one-question, one-answer paradigm doesn't leave a whole lot of room for flexibility in style. This may be why most kids seem to feel most truly alive outside school, participating in extracurricular activities that not only allow for recognition of individual differences but also insist on it. The high school soccer coach doesn't start with a view that every player is just as good in goal. Team members are assessed based on their individual qualities and then slotted in where they fit best. Once identified, unique strengths are emphasized and developed.

Growing up, my own extracurricular was, to my parents' chagrin, horseback riding. They found a way to support it financially (though I was never in the league of riders with first-class horses, instead riding other people's, it was still an expensive pursuit, and I'm sure they thanked their lucky stars that only one of their seven kids was interested). Horseback riding is all about thinking processes—yours and those of the somewhat dim-witted but powerful animal you hope to control. You need to understand how a particular horse thinks if you have any hope of getting it to do what you want, and you need to be fairly methodical about how you make commands and when. I rode dressage, a somewhat technical form of riding where the entire point is to get responses from your horse using minimal signals; ideally, an observer can't even perceive the signals. You can't do it well unless you understand and reach the horse on another level, which is what I loved about it. You're forced to bend your mind to see the world the way a horse does, and that changes the way you think, too.

For whatever reason, it occurs to us less frequently that we should be trying to do the same thing with people. Yes, we understand that other people have different histories, agendas, points of view and temperaments, but often we just don't recognize or account for the fact that their brains process information differently than ours do, too. I'm not talking about intelligence but literally about the way we think, which is as individual as we are (and, as we'll see in a minute, this has some pretty important implications for how we interact with others).

How do I know this with complete certainty? Because I am an identical twin. Literally clones, identical twins are freaks of nature. Unlike fraternal twins, whereby two eggs happen to be in the right place at the right time and are both fertilized, identical twins, for reasons that are still mysterious, started as one fertilized egg, then split in two. So my sister Adrian and I started out, genetically, exactly the same.

Yet even in the womb our development diverged. By the time we were born, I was a full pound heavier and an inch longer. As we grew up, other differences emerged. She was a thinker; I was an emoter. She was socially gifted; I was shy, bordering on awkward. And so on. We were pretty much equally smart (depending on the subject). Our Grade 10 computer science teacher even made us retake the final exam because she couldn't figure out how we'd managed to score identically even though we'd been sitting across the gym from each other. But do we think alike? No way. We process things very differently, even if in the end we arrive at the same conclusion. From the time we were little kids, the way we saw the world and interpreted information was just not the same.

One of the first questions I had when I started to think about curiosity and innovation was why some people are innovative and open to change, and others are not. I wondered if, maybe, some people just aren't capable of innovation. All the experts told me that isn't the case: everyone can innovate, we just do it in different ways. For some people, innovation involves concocting a new kind of martini. For others, it involves coming up with a new app. We're all capable of it to some degree, but we do it differently, in large part because our brains don't work the same.

Few people understand this better than Robert Rosenfeld, co-author of *The Invisible Element.* He is frequently called on to help companies become more innovative. In his work with clients over the years he has relied on various personality inventories such as the Myers-Briggs. But he was frustrated by them because they didn't reveal "the complete story or even the essence of the problem" when it came to innovation. Existing tests didn't focus tightly on the connection between style of thinking and openness to innovation, nor did they have the scope to measure the full spectrum of ways that people think about and approach problems. So he created his own test: the Innovation Strengths Preference Indicator (ISPI).

The ISPI is one of the only tests that looks at both your risk profile and your approach to implementing change, which is more about outcomes. The test gauges ideation (idea generation), risk taking, and process (how you go about implementation), and ranks you in those three domains: you can be an extreme builder, mid-builder or builder, or an extreme pioneer, mid-pioneer or pioneer. Builder types like to work with what's already there and adapt it incrementally to make it better, whereas pioneers prefer to forge something new and untested. The ISPI also measures how you relate to others, whether you need to be in control, the extent to which you initiate relationships and your networking style. Finally, it looks at aspects of your personality, such as passion and energy. Because there are in total twelve categories (most personality tests cover three or four) and you have an individual score in each, there are 38 million unique variations of the ISPI.

I think the trick with tests like this is to do them fast so you're sure to write down your first, instinctive response rather than the one you think sounds right or better. I took the ISPI—basically a series of questions about how you approach problems and other people—in about fifteen minutes one cloudy winter afternoon, sitting at my desk at the CBC. Afterward, Rosenfeld's colleague Andrew Harrison walked me through the results.

Unlike personality tests that peg you as a definitive type, the ISPI tells you what type you are in all three domains related to problem solving and innovation. It reinforced some things I already knew to be true about myself but also helped me understand something important about how I approach problems. I like new ideas, but I want to use them in practical ways. I've always thought of myself as very curious and open to change, but the truth is, I'm just more comfortable seeking to improve what already exists.

According to the ISPI, I am a mid-builder when it comes to ideation, a pioneer when it comes to risk taking and a mid-pioneer when it comes to process, meaning I'm capable of some flexibility in terms of how things are implemented. I like to dive right in to try to fix things, so I'm a first responder in a crisis, but I may not have a lot of original ideas about how to handle it (and I am more likely to succumb to the allure of old assumptions). This is information I could really have used back in university because it explains perfectly why I could never be an innovative architect—and also why journalism, where I'm reporting on and interpreting events and theories, is such a good fit for me. I don't have to come up with breathtaking new ideas all the time, but I do need to be willing to go out on a limb and also need to be flexible about how I find material and weave it together. I can be innovative, just in a way different from how I hoped and dreamed when I was nine years old.

It's pretty obvious that this kind of self-knowledge isn't helpful just professionally. It can also be helpful when it comes to trying to innovate and solve problems in your personal life. Even if you don't know where a colleague or friend places on the ISPI, you know there is a place, and it's probably not the same as yours. Acknowledging that can really change how you perceive others because now you have a new way to look at their behaviour when you're facing a problem or having a conflict of some sort. Innovation is required to fix the issue and move on. If someone is frustrating you, perhaps it's not because she's deliberately trying to thwart you—it isn't about you at all. Maybe she is just proceeding in a certain way that accords with her place on the ISPI spectrum. If your own is far away—she's an extreme pioneer but you're a mid-builder, say, so her desire to challenge the status quo bumps up against your tendency to focus on improving what already exists—well, no wonder there's conflict.

It's liberating, soothing even, to understand that friction isn't necessarily a verdict on your personality. Just as you have no real idea what someone else's headache feels like, you really have no clue how his or her brain processes information. In almost the same way that emotional maturity involves growing out of childhood egotism, intellectual maturity might involve realizing that ours is not the only way of thinking. Recognizing that we all approach problems differently can help you to be more patient and empathetic, and will probably help everyone reach a solution faster. And it also helps to recognize the incredible value of being exposed to different ways of thinking, whether they were acquired living abroad or living right next door.

For me, it's been humbling yet freeing to recognize with real clarity that my way isn't necessarily the right one. It's just my way.

CHAPTER ELEVEN

Don't Stop Thinking about Tomorrow

Curiosity is the antidote to complacency, but only if you act on whatever it is you discover. Many floundering companies don't get in trouble because they didn't ask questions but because they didn't believe or weren't willing to embrace the answers they found. Instead, they relied on past experience: the products, services and approaches that had always been winners in the past.

This is what appears to have happened at Microsoft, which was once one of the most innovative companies in the world and now is regularly eclipsed by Apple. The reason, according to some company insiders, is not that Microsoft isn't coming up with great ideas, but that those great ideas are being sabotaged internally. Now, the company is not exactly hurting. It remains hugely profitable, but the vast share of its profits come from Windows and Office programs that were created in the dawn of the PC age. While Apple has wheeled out one cool, beautifully designed piece of hardware after another—the iPod, the iPhone, the iPad—Microsoft has never been able to gain traction in terms of hardware and has lost market share in smartphones and fancy laptops. (Microsoft *is* venturing into the tablet market, to be fair, but history suggests it may have an uphill fight.) The issue is not that Microsoft is only good at software and can't come up with cool hardware and technology, according to Dick Brass, who was a VP at the company for seven years. The issue is that visionary new products were routinely pushed to the sidelines because of what he calls "internecine warfare" within Microsoft. Office politics, in other words.

For example, graphic experts invented ClearType, a way to make type more readable. It was developed to help sell e-books but could've been a competitive advantage for all devices with screens. It unleashed competition all right—within the company itself. "Engineers in the Windows group falsely claimed it made the display go haywire when certain colors were used. The head of Office products said it was fuzzy and gave him headaches. The vice president for pocket devices was blunter: he'd support ClearType and use it, but only if I transferred the program and the programmers to his control," Brass explained in an op-ed in the *New York Times*. "As a result, even though it received much public praise, internal promotion and patents, a decade passed before a fully operational version of ClearType finally made it into Windows."

Internal competition can promote innovation, but not if what's being contested is turf rather than ideas. Brass noted that one hardware innovation after another had failed to launch because of turf wars that created "a dysfunctional corporate culture in which the big established groups are allowed to prey upon emerging teams, belittle their efforts, compete unfairly against them for resources, and over time hector them out of existence. It's not an accident that almost all the executives in charge of Microsoft's music, e-book, phone, online, search and tablet efforts over the past decade have left." Microsoft has an enviable past, he concluded, and a prosperous present. But its future prospects are less than bright.

A "why mess with success" attitude leaves businesses vulnerable when the environment changes. Relying on what's worked in the past can backfire spectacularly. Consider General Motors. In 2008 and 2009, as the North American economy was in free fall, the company chose to maintain its focus on big, gas-guzzling SUVs. Sticking with the status quo seemed like a perfectly reasonable choice. "All the data pointed that way," says innovation expert and Rotman dean Roger Martin. "Sales of those vehicles, based on past experience, trended straight up. Who knew there would be an inflection point where the line would plummet dramatically?"

Certainly not GM, whose leaders had never felt the need to question their assumptions. After all, GM was the world's largest automaker, with a collective corporate ego to match. When then CEO Rick Wagoner first went to the U.S. Congress in October 2008 to ask for a financial bailout, he testified, "What exposes us to failure now is not our product lineup, or our business plan, or our long-term strategy." The only problem, Wagoner insisted, was the recession.

In reality, the company's problem was that, like Microsoft, GM had a proud history of fostering innovation—then smothering it. For instance, the company's engineers came up with minivans in the 1970s, a decade before Chrysler did, and with the EV1 electrical vehicle in the 1990s, but in both cases, senior executives wouldn't commit to following through. They focused instead on the bottom line and what had always worked before, rather than looking to the future—a strategy that ultimately landed them in Washington, DC, cap in hand, begging for a handout.

Ford, on the other hand, made a different set of choices, and in 2010, the company had one of the most profitable quarters in its history, an extraordinary achievement when you consider the economic backdrop. Roughly 18 million Americans were still out of work, and car dealerships were floundering because people didn't have the cash to buy new vehicles. Furthermore, of the three big North American automakers (Chrysler is the other), Ford was the only one *not* to be awarded a government bailout in 2009. Partly that was luck. The company had, coincidentally, raised a bunch of money just before the recession struck and was therefore flush, relatively speaking. Which might sound great, except that it put Ford at a slight disadvantage; its competitors were able to extract concessions from workers, who feared for their jobs.

So how did Ford manage to pull off a stellar quarter? By embarking on a strategy that was a daring departure from past experience. Up until 2010, the company's top seller had been the F-150, a big gas guzzler of a pickup truck. But unlike GM, which opted to stay with the tried-

and-true post-2008, Ford looked at the consistently high price of gasoline (even through the recession, gas prices were stubbornly high) and decided fuel efficiency should be its target. And that meant smaller vehicles. So Ford turned its attention—and marketing clout—to smaller cars like the Focus and the Fiesta.

Nothing in its business model or past experience ensured this was the right call. Ford had to make a conscious decision to correct for the status quo bias and go with a different approach, one that was informed by projections about the future. But Ford's leaders didn't just cross their fingers and hope for the best. They controlled what they could, creating ad campaigns to promote the vehicles and adding all kinds of bells and whistles to jazz up the lower-end cars. Innovation didn't require inventing a whole new kind of vehicle, but rather, simply introducing optional features that were standard in luxury sedans—GPS systems, docks for handheld devices, heated seats—but not common in smaller economy cars. Those extras aren't just appealing optional add-ons for consumers; they're also relatively cheap-to-manufacture, high-margin items for automakers. The Fiesta quickly became a top seller for the company, and Ford's profit per car doubled from $1,000 to $2,000.

Interesting, isn't it, how easy it was for Ford to innovate? All it really had to do was borrow some features from other products it already manufactured. The hard work, when you think about it, wasn't finding answers. It was having the courage to implement them.

Loss aversion, and the temptation to rely on traditional methods and approaches, is most acute for public companies, which must meet quarterly targets to keep analysts optimistic and shareholders happy. Experimentation and innovation can simply feel too risky. For years, Roger Martin points out, General Electric's flagship product "wasn't an industrial turbine or refrigerator or medical imaging device, but a quarterly earnings number that reliably met or ever so slightly exceeded" analysts' expectations. The danger of thinking this way is clearest in a time of heightened fear and emotion. At exactly the moment a business can least afford it, leaders may, instead of questioning their basic strategy and assumptions, insist on them even more adamantly. Often the perspective is "We already *know* what's wrong; we're losing money. We need to focus on solutions already. And doing x is what's worked for us in the past."

Looking at failed businesses, in retrospect it almost always seems that there was a critical window of opportunity when disaster could have been averted. But the people in charge stubbornly insisted on hewing to the path they'd always taken, and simply didn't ask the right questions—or in some cases, any questions at all—thereby running the risk that they'd become the modern-day equivalent of the last buggy whip manufacturers.

One of the best and saddest examples is Kodak, which for generations was the global leader in the development of photographic images. As unlikely as it seems now that the company is in bankruptcy, Kodak was also one of the most innovative players in its industry. Invention was part of the corporate DNA. Back in 1880, George Eastman, the company's founder, loved still photography but was frustrated that cameras were so heavy and bulky. He quit his job as a banker and moved to London, then a centre in the world of photography, with the goal of making a more user-friendly camera. In 1884 he patented the process of manufacturing rolls of photographic film and spent years developing a camera to take advantage of it. The first Kodak cameras were expensive—$25 apiece, a fortune at that time—and were sold loaded with film. To have it developed (cost: $10), you shipped the whole camera back to the factory, where a new roll of film was inserted while the old was printed into photos. This was also a hugely profitable business. But Eastman, one of the most generous and unassuming philanthropists of the day, wasn't satisfied with that outcome and kept working on creating a more accessible and affordable product.

In 1900, the company launched the Brownie, which retailed for a dollar, and by 1930 Kodak was included in the Dow Jones Industrial Average index, which tracks the fortunes of a small group of companies chosen to mirror overall trends in the U.S. economy. As other companies entered the camera business, Kodak focused on improving film, thereby insuring that its products would be used even in competitors' cameras. In 1932, Eastman, suffering from the same degenerative disorder that had crippled his mother, committed suicide, but by that point the company was the established leader in its industry.

And then, a major opportunity missed. In 1975, a Kodak

engineer came up with a heavy, cumbersome device that would expose an image after twenty-three seconds—one of the first working digital cameras. At that time, the company owned 90 percent of the American market for film and had an 85 percent share of camera sales. Perhaps this dominance bred complacency; certainly it created a strong incentive to continue focusing on film. After all, everyone would always need film, right? In any event, Kodak didn't develop and exploit its own digital device. Its half-hearted approach to the digital market was underscored in 1994 when it launched a consumer digital camera in Japan under its own name but then allowed Apple to license its technology and launch a version in the American market called QuickTake. Just ten years later, Kodak's traditional camera business was shuttered, and the company was never able to stay far enough ahead of the technological curve to gain real traction in the digital camera space—and, of course, the market for film cratered. In 2012, the company filed for protection from its creditors.

One of many bittersweet ironies is that, in bankruptcy, Kodak's most valuable assets were some of the digital patents it owned but had failed to exploit properly. At several points on the road to going broke, the company appears to have chosen to ignore new trends, even when it was perfectly positioned to compete or even lead. In the end, relying on old assumptions—and abandoning George Eastman's tradition of continuous innovation—proved to be its undoing. The most fundamental questions a company can ask—"Who are we and what do we sell?"—might have helped Kodak refocus and regroup. If its answer had been "we sell memories" instead of "we sell film," it might have hopped on the digital bandwagon and steered itself toward a different kind of future.

HOW CAN WE AVOID A KODAK MOMENT IN OUR OWN LIVES?

Many of us make the same mistakes that GM, Microsoft and Kodak did. We come up with good solutions to our own problems, ways that we could innovate to improve our future, then fail to implement them. How can we make ourselves follow curiosity to its logical conclusion and embrace answers that take us away from the status quo?

A friend who's a management consultant makes a lot of money helping clients do exactly that. But when it came to her own life, she found it much more difficult. Recently divorced, she had known for almost ten years that she should leave her marriage. The reason was straightforward. She not only didn't love her husband, she didn't like him either. "I look back at my own reasons for failing to act and they seem really flimsy," she says today. For years, she told herself she couldn't leave because she wanted children; when she and her husband didn't have kids, she told herself she couldn't leave because she didn't want to be alone.

The real problem, she says, was fear of the unknown. "In retrospect, it was ridiculous," she says. "I wasn't a victim of abuse, I wasn't unable to support myself, I wasn't clinically depressed or incapacitated in some way, I don't belong to a religion or come from a culture where divorce is frowned upon." She really had no good reason to stay other than the status quo bias.

How, then, did she wind up making herself leave? First, she asked herself what, specifically, she feared about divorce. "I made up a long list of things, I tried to be as specific as I could, then dealt with them one by one. For instance, I was worried about having nothing to do on the weekends, so I joined a book club and signed up for tennis lessons. I told myself, 'You're not committing yourself to divorce, just to some weekend lessons.'" But by forcing herself to question, in a very concrete and detailed way, how she'd implement her solution—divorce—she convinced herself that it was the right answer and embraced it.

Second, she questioned what the future would be like if she stayed where she was, and tried to be just as concrete and detailed as with her first question. "I figured out what my life would look like in five years, ten years, fifteen years if I stayed. I thought about things like what would happen if I got really sick, or if he did, or if one of us lost our jobs, or one of our parents needed to move in—I really tried to think not about how things were at that moment but about what *could* happen and how I'd cope or feel. I did exactly what I help big companies do when they're coming up with long-term strategic plans, and I tried to be just as rigorous and objective."

She is, by the way, happily single, with just one regret: that it took her so long to do what she knew was the right thing.

NEW ISN'T ALWAYS ENOUGH

Some companies and individuals are fully committed to implementing forward-looking solutions they've dreamed up, but nevertheless stumble because they rely on an old delivery model, then can't gain traction. For instance, providing dental care to underprivileged families in Charlotte, North Carolina, seemed like a straightforward and worthwhile goal to Dr. Ryan Root. He expected people to line up for free cleanings and exams at his clinic, since even basic dental hygiene is beyond the means of a lot of the working poor. So he was surprised when a huge proportion of the low-income clients who eagerly signed up for free care then missed their scheduled appointments.

Root could have concluded that poor people are perhaps not educated enough, not considerate enough, perhaps simply not—*insert stereotype here*—enough to bother keeping their appointments. Instead, he probed a little more deeply. Why were poor people standing him up, again and again? What exactly was the problem? Not a lack of desire for free dental care, he discovered. The issue was something that had never even occurred to him: the bus fare to get to his clinic was, for patients who really needed free services, a deal breaker. Once the dentist began providing bus tickets, people began showing up for appointments like clockwork, more or less.

If you're trying to strike out in a bold new direction but using the same old road to get there, you may not ever arrive. As cautionary tales go about what happens when you get the product right but the delivery model wrong, the Tata Nano has to be high on the list. The Nano is the dearly beloved creation of Ratan Tata, head of the Indian conglomerate that bears his name. In a country where the most common mode of motorized transportation (for those who can afford it) is a two-wheeled scooter, he envisioned a small car that would be affordable to millions of Indians, newly minted members of a growing middle class. Technically, the Nano is a feat of engineering. Lighter than other cars, it has fewer moving parts, which makes it far less expensive to manufacture. The trunk, for instance, doesn't open from the outside and can be accessed only from inside the car—a relatively small inconvenience if your previous vehicle's "trunk" was your own back. Also appealing was the price tag, about $2,500.

The Nano is the cheapest car in the world, and Tata fully expected it to revolutionize travel in India. When the car was officially launched in 2009, the company geared up to ship 25,000 a month. There was no need. Two years later, the company was still selling fewer than 2,500 a month. The problem wasn't the Nano. The problem was the way it was being sold.

The end goal—to provide cars to people who'd never been able to afford them before—got lost in the shuffle. Tata's first mistake was to market the Nano through showrooms in big cities, so its core target market in smaller towns didn't have access to the car. And those who did soon discovered that it was difficult to arrange financing. Yes, Tata had made a very inexpensive car, but the company approached financing the same old way automakers always had. So strict were the lending criteria that those who passed the test and qualified for a loan quickly realized they could also qualify for a bigger loan and get a "real" car. Those who would have been delighted to drive a Nano off the lot were sent packing. This helped solidify the public perception that the Nano is "the poor man's car," not something to aspire to but something to settle for. That just about killed the prospects for Tata's incredible innovation.

But before that happened, the company stepped back and questioned its reliance on an old model of auto sales, which was a terrible mismatch given its product and its customers. Today, Tata is opening dealerships in small towns, buyers can put just a few hundred dollars down and drive off in a Nano, and financing is more easily arranged with local banks. The moral of the story? It doesn't matter how worthy your goal or how innovative your idea is. How you deliver it will usually require innovation, too.

The same is true in everyday life. We come up with a visionary goal for ourselves, know exactly where we want to go, but flounder en route. Very often, this is because we make the same kind of mistakes Tata did. We rely on conventional wisdom or an old way of doing something—or what has worked well for others—when we're trying to achieve something new. And then, when it doesn't work, instead of questioning how we're going about it, we decide the problem is either personal incompetence ("I just can't do it, I'm not up to the task") or the goal itself ("Being a manager is overrated").

Having a new goal or idea is highly energizing: I'm going to learn Spanish before I go to Mexico! I'm going to learn to cook—for real this time! But actually achieving big goals in our everyday lives isn't quite so exciting. It's more like Addison Lawrence slogging away in the lab, taking water samples and concocting smelly shrimp foods. Implementation requires a certain bloody-mindedness, a willingness to keep questioning and trying to figure out a new way when you encounter an obstacle or your willpower flags. And in order to do that, you have to find ways to keep yourself engaged and curious, and that's not always easy.

I know this because I like to run. Correction: I like to *have* run. I'm not one of those people who enjoys physical exercise for its own sake, which is why running suits me. It's efficient: I can burn a whole lot of calories in a relatively short period. I'd rather burn calories sitting on a couch with a great novel, but that hasn't worked out for me. To make it more enjoyable, I run with my dog, a chocolate lab named Bella, and whenever possible, my sister Adrian. She is a better and more disciplined runner than I am, but she's willing to slow down so we can spend some time together chatting, and her company keeps me engaged and running, even on mornings when I'd love to sleep in.

One of our regular routes takes us through a ravine, at the end of which is a big hill. I'm not being a wimp, this is a serious hill—about 300 yards high, at a forty-five-degree incline. And for the past three years, my goal has been to be able to run up it without stopping. The cardio and conditioning benefits of hill running are incredible, and apart from that, I feel like being able to make it up that incline would mean I'm a real runner, like my sister.

But I can't do it. At about the halfway point, it's always the same story. My lungs are burning and my heart is hammering and I have to stop. Meanwhile, Adrian runs on steadily, my loyal dog trotting beside her without so much as a backwards glance at her owner, who may or may not be in cardiac distress.

Okay, so the hill is a challenge for me. But I like a challenge. Keep running it, and I should keep building my endurance and conditioning, and eventually make it to the top. In theory. In reality, no matter what changes I've made—attacking the hill the way Adrian does, keeping a slow and steady pace the way I do on other hills, varying my speed the way running magazines tell you to—I've never been able to get past the midway point. Until a few months ago, that is, when Bella was, uncharacteristically, lagging behind so I had to stop at the bottom of the hill and wait for her while Adrian steamed up it at her usual pace. And a funny thing happened. It was only about a forty-five-second pause, but my heart rate slowed, my breathing calmed and when I started up the hill I felt great, passed the dreaded midpoint and just kept going. And for the first time, I made it all the way to the top without stopping.

Wait a minute, you might say, you *did* stop, at the bottom. And it's true, I did. But once I'd done that, I thought about it. My goal is to reap the cardiovascular and conditioning rewards of getting up that hill at a good clip. A little innovation, thanks to my not-so-clever dog, makes it possible for me to achieve it. Both the right way and my old way just didn't work for me. But I'd stuck stubbornly with the tried-and-true approach despite the fact that it never succeeded (though I probably did burn off a few calories beating myself up for my failure).

And in a weird way, I'd disengaged and stopped questioning. My body was moving, but my mind was on autopilot, certain that one day the traditional approach would magically work because it just made so much *sense.* To paraphrase management expert and author Gary Hamel, I'd become the prisoner of my own expectations. Just as some businesses do when they're trying to reach new goals or create new markets, I'd burdened myself with an outmoded set of expectations that actually limited my possibilities.

LOOKING TO THE FUTURE

As we've seen, we all have the capacity to innovate. Think of how kids in developing countries respond to the Soccket. When they see the ball, they almost immediately start asking questions and dreaming up their own innovations (at least once they've stopped kicking it around). Innovative thinking is contagious, which is one good reason to surround yourself with people who think differently and have had different experiences from you. Innovation is like a bug that anyone can catch.

But you have to *want* to catch it. Many of us prefer to wallow in the present, especially when everything seems to be going along smoothly. Forcing yourself to consider new possibilities, whether at work or at home, requires not only discipline but also some fundamental belief in what Rolf Smith calls Me, Inc. Without that, you may lack the conviction to act. It helps, too, to give yourself permission to make a few mistakes, as Addison Lawrence does—and to remind yourself to analyze them so you learn something from them. A playful approach, even when the stakes are high and the cause is important, is also a plus. Immersing yourself in a problem and really playing with it, as Sean Moore and Steve Gass did, helps you stay engaged.

But there's really only one essential ingredient when it comes to innovation, and by now you know what it is and that you already possess it. Curiosity comes naturally. If you're able to keep it alive, stoke it and then follow it where it takes you, you will find success—however you measure it.

CONCLUSION

Imagine you could design a university from the ground up. Not just the campus and the buildings, but also the classrooms and what happens inside them. What would the place be like? What would students do there, and what would they take away?

The first thing you might do is scrap the model of instruction that's prevailed for hundreds of years, wherein a professor lectures at the front of the room and students take notes. That model is, after all, an artifact of a time when printed material was scarce, so an instructor's job was primarily to tell students about the book he'd read (or written). Since you're changing all that anyway, you might also want to toss out the notion that information should flow in just one direction. You might be swayed by all the new research on how we learn and process information, and opt for highly interactive courses, where expert guides are standing by but students are also allowed to explore and make discoveries.

You might design a curriculum that aims to get students to drill down, instead of skimming along the surface, and to that end, perhaps you'd decide to organize things so that students could focus on just one subject at a time rather than juggling five or six. You might, reflecting on your own experiences and all the things you "learned" but didn't retain because you weren't really all that interested in the first place, let students shape their own curriculum to some extent. You might even decide *they* should be the ones asking questions, and that those questions should dictate their academic path because, at the very least, they'd be studying things they cared about. Of course, you'd want to make sure they covered some required basics so they wouldn't wind up with glaring knowledge gaps, but how would they do that exactly? Maybe, knowing that people think and learn differently, you'd leave that up to them.

When you do turn to designing the buildings, you probably wouldn't opt for cavernous, anonymous lecture halls. Maybe you'd go for something more like what you have at work: informal seminar rooms where small groups could sit around a table so that it was more likely students would interact with each other and collaborate with, rather than listen to, their professor. The spaces might have clean lines and huge windows so that people could feel connected to the outside world, and at night, lit from within, the buildings with their clear glass facades would look like jewel boxes. If you were really planning, you'd also make sure that that outside world included snowcapped mountains, great swaths of green, and cool mountain rivers providing staggeringly beautiful outside labs.

The real point of a university, you might decide when thinking about how to build one from scratch, isn't to hand out degrees. The point is to encourage and foster certain traits, like curiosity, so that students want to keep exploring and learning even after they leave. Your hope would be that by then, they'd be fearlessly curious and hooked on learning—not only because that would make their lives better but also because it would be a pretty big contribution your university could make to society: cranking out innovative thinkers.

David Helfand was an astronomy professor at Columbia University in New York City when, in 2005, he was asked to do exactly this: reimagine higher learning for the twenty-first century. And the result—the innovative place I've just described—actually exists. Nestled in a valley at the foot of the Tantalus mountain range in British Columbia, Quest University aims to teach young adults to think for themselves.

Helfand, who initially came to consult on the project, found he couldn't stay away. Now Quest's president, he has a disarmingly youthful voice, a bushy beard that springs out from his jaw and bottomless enthusiasm about what the university is attempting and accomplishing. For him, the opportunity to create a truly unique post-secondary institution couldn't have come at a better time; Helfand had become disillusioned with the Ivy League.

Columbia, one of the most venerable of the Ivies, had become, as far as he was concerned, just a degree factory, a place that spent millions on recruiting students to apply, then turned most of them away. It had become a game. The prestige and worth of a school are increasingly measured not by how inspirational its teachers are or by how far it stretches students' minds, but by how difficult it is to gain admission in the first place.

His awakening—though at the time, it felt more like a blow—occurred on the day a laconic freshman informed him, "We aren't here for an *education.* We're here for the degree." It was a disappointing moment for Helfand, a somewhat idealistic maverick who'd refused tenure because he worried it might lead to intellectual complacency. But in truth, his disenchantment had started years before.

When he arrived at Columbia as a young professor, he'd been pleased to discover that core courses were mandatory in the first year; regardless of what a student majored in, there were some ideas the university insisted were essential knowledge. Helfand was less pleased to discover that all seven required courses were in the liberal arts and humanities. Not a single math or science course. "Well great, I'll just add some science and math," he thought innocently. Twenty-seven years later, he'd succeeded in adding one survey course, his own Frontiers of Science. Meanwhile, it had become undeniable that most of the young people he taught viewed university primarily as one stage of a "hurdle race" to the workforce. Wide-ranging curiosity and intellectual growth were not on the agenda.

What brought this home to him, with sickening clarity, was an afternoon he spent with some fourth graders at a small school in Manhattan. At the end of his thirty-minute talk about space, every small hand in the class was up in the air. Forty-five minutes later, hands were still up and the kids, still calling out questions, had to be dragged out of the room. Energized by their enthusiasm, Helfand headed back uptown to Columbia to teach his Frontiers of Science course. Twenty students were waiting for him in the seminar room, dozing, busily texting or simply gazing blankly ahead. "Why can't you be more like fourth graders?" Helfand blurted out. Ignoring the rhetorical nature of the question, his students informed him that whereas fourth graders have a lot of questions and like to learn, they just needed to pass his course so they could get degrees so they could get high-paying jobs.

Shortly thereafter, Helfand took a leave from Columbia and headed to British Columbia to try to come up with something really innovative: a university where students could be—had to be—more like fourth graders.

ONE THING AT A TIME

Quest's curriculum is organized on a block system, which means that students study one subject—and one subject only—for three and a half weeks straight. They are in seminars three hours a day, interacting with the professor and each other, and then spend another five hours working outside the classroom. The beauty of the block, says Helfand, is that it's tailored to the way our brains absorb information; as a species, we just aren't very good at multi-tasking.

Quest has core courses—sixteen of them, spread out over the first two years—but the emphasis is on learning how questions are asked and answered in different fields. "The point is not to deliver content or a particular skill set," Helfand explains. "The point of our core courses is to show a student how a physicist asks questions about the world and tries to answer them, how a philosopher does that, how an economist does that. And so the content—the amount, and even the nature of it—is secondary."

The focus on just one subject at a time also allows professors to be as creative as they want. Spur-of-the-moment field trip? No problem. They're encouraged to try out new ways of teaching material, too, and Helfand himself has found this helpful. At Columbia, for instance, he taught a class on planets and always started by writing down Kepler's laws, which describe how planets move around the sun. At Quest, he doesn't teach the students Kepler's three laws but instead shows them an advanced simulation and then guides them through a series of questions. In the end, the kids themselves figure out the laws. "They derive Kepler's laws in the same manner Kepler did," Helfand sums up. "It's the Confucian thing: 'If you tell me, I'll forget it; if you show me, I'll remember it; if you involve me, I'll understand it.'" In the block system, professors have the luxury of time, and time is what you need if you want to get students engaged and involved and figuring things out for themselves.

Second-year Quest student Maymie Tegart attended a pretty typical high school in Kamloops, British Columbia, where she was one of 1,300 kids. A straight-A student with a full slate of extracurricular activities, Tegart nevertheless didn't feel she was learning a whole lot. "I was never expected to actually *think* in my classes" is how she explains it. "It was a lot of memorization, a lot of just sort of getting us to the diploma."

But she *wanted* to learn to think, and therefore concluded that a traditional university might not be the way to go. It might be a little like her high school. Still, she worried about gambling on a brand-new institution. "My first question in the interview," she remembers, "was, 'How do I know you're going to survive?'" The answer she got was convincing enough that she decided to take the risk. And she certainly has been learning to think.

By the end of first year, all Quest students are expected to have come up with a question that will guide and shape the rest of their coursework. Tegart's is about food production and safety. That's not what she came to university intending to study, but as she reflected on what she was really curious about, she realized that, unbidden, she'd been focusing academically on food; her projects had ranged from a rhetorical essay on bug eating to a molecular biology project examining stem cells in meat production. And so, today, she is designing her coursework to satisfy curiosity about a subject she wasn't even consciously aware she was interested in until she was allowed to explore intellectually.

Early on, students at Quest are taught the value of collaboration by heading out in the woods for a team-building exercise, the kind many academics might dismiss as touchy-feely hogwash. Helfand certainly did. "Being a classic East Coast Ivy League intellectual, my eyes almost got stuck in the top of my head when I realized we were going to give up almost two days of academic courses to this nonsense in the woods," he says. "But I would never change it now.

The difference it makes in the classroom, when they come back from this two-day excursion, is remarkable. They understand what it means to work together, to draw on each other's strengths, cover each other's weaknesses."

Quest doesn't think students are the only ones who should embrace new ways of doing things. For instance, its first-year "cornerstone" courses were developed to ease students into the block system and also ensure they had basic presentation, reading, and writing skills; after five years, the faculty members who teach cornerstone courses had them down cold. And this was a concern. What if they started to dial it in, rather than engaging fully, and the courses started to feel stale? So there's now a committee looking at redesigning the cornerstone curriculum. Its name? The Committee to Blow Up Cornerstone. New material and new approaches will presumably be introduced by staff, who will be re-energized, and then, in a few years, when it has that same old, same old feeling—well, they'll blow up cornerstone again and go back to the drawing board.

The staff's openness to new experiences and new approaches, says Tegart, jump-starts students' curiosity. Now, she says, she takes nothing for granted. "If people tell me something, I'm always asking how they know it's true, what the source is. When I hear something and have an opinion, now I also think, 'But it could also be this way, or that way.' I'm always trying to draw new conclusions."

That's exactly what Quest wants students to do. "I hope what we're arming students with, or drawing out of them, is curiosity," says Helfand. "Intellectual fearlessness. Just ask the question. You won't look stupid—just ask."

Consider what the rest of us can learn from Quest. Yes, these are kids—in their early twenties, for the most part—and therefore probably a bit more open to change than they will be even a few years after they graduate. But the curiosity that is so essential to their success at Quest is not some rare gift. Once upon a time we all had it, and we relied on it not just to acquire information but for sheer pleasure. Being curious is the same as being interested. And being interested feels good.

Perhaps there is a way to redesign our own mental environments in precisely the way Quest is redesigning the university experience, so that we're more interested and feel more alive and engaged. Perhaps, like Maymie Tegart, we can all start to think of ourselves as innovators and challengers. To do that, though, we need to blow up some of our own cornerstones, our myths about what innovation actually is.

Myth #1: Innovation Is about the Newest Thing

Sometimes a great innovation is indeed a "step-change": the motorized vehicle that displaces the horse and buggy. But most innovation is incremental. From my own favourite, life-improving innovation—the curved shower rod—to just about any product or service you can name, little improvements and developments are being introduced all the time. If you're using an electric razor that is more than five years old, for God's sake, run out and buy a new one. You'll be shocked and delighted by how much better the technology is these days. Even something as basic as a hammer is always being improved upon—making the grip more comfortable, the steel alloy lighter, the range of sizes greater so that even if you're not a burly six-footer, you can find one that actually fits in your hand.

Of course, the desire to keep improving products, processes and services is partly driven by a less-than-altruistic question: How can we make more money on this thing? But there's also a natural human instinct at work and another kind of question: Can I make this better, just for the intrinsic satisfaction that comes from improving things? This is a question we can and should ask ourselves a lot more often, not just at work but also in our personal lives. If there's something you've been doing exactly the same way for a long time, whether it's how you clean the house or the kinds of conversations you have at the dinner table, there may be incremental improvements you could make that would have a major payoff in terms of your overall sense of well-being.

Myth #2: Innovation Is a Solo Activity

Consistent with our tendency to think of innovation solely in terms of mind-blowing new inventions, we often think of innovators as geniuses, oddballs with wild ideas and wilder hair. And some of them do fit the stereotype. People who occupy the far end of the innovation spectrum were probably less easily tamed by our school systems and may therefore be less comfortable in corporate environments. But even mavericks and mad scientist types need other people to implement the innovations they've dreamed up, and usually, those other people wind up incrementally improving their inventions in some way. Remember Steve Gass of SawStop? It's true that his moment of inspiration came when he was alone—and he even took the next step, working out the math to see if a safer saw really was possible, on his own. But he was quickly joined by two colleagues from his law firm, people so inclined to challenge others' ideas that a visitor to the office was taken aback. What turned Gass's jerry-rigged innovation into an actual, functioning business was collaborative energy. The three men were able to bounce ideas off each other and figure out, together, how to run a business and how to manufacture a product, and then keep right on improving it, incrementally.

Myth #3: Innovation Can't Be Taught

Every day, people like Colonel Rolf Smith teach organizations, businesses and individuals how to get in touch with their inner innovator. But teaching innovative thinking isn't like teaching math or French—it's more a matter of teaching people how to harness their existing natural curiosity in order to unleash their innate capacity for innovation. Think of Smith's rock-climbing exercise. The point of it is to take someone who thinks, "I can't rock climb—I don't have the skills, and it's too risky," and let him show himself, "I can rock climb! It's a bit risky, but I already have some skills I didn't know I had, so I can figure this out." That's also the point of Me, Inc. Simply by getting to know yourself and recognizing your strengths and weaknesses, as well as the opportunities and threats you're facing, you'll perceive your capacity to innovate. We also know innovation can be taught because of places like Quest, where students are taught both explicitly and implicitly that curiosity launches all positive change—and, by the way, it's intrinsically rewarding.

Myth #4: Innovation Is Top-Down

Remember the flocking theory? Flying in formation, birds on the periphery—where the risks are, and where you can see more—send messages and warning signals to birds flying in the centre, where it's more protected and safer. Similarly, in a fast-food restaurant, the clerk at the counter cottons on long before anyone at head office does that the new trays are flimsy and hard to stack. In a hospital, the nurses may resist washing their hands unless there's a way to communicate both up and down the food chain that the problem is the harsh cleanser they're made to use. Smart companies like Four Seasons and Whole Foods explicitly recognize that the closer an employee is to the end-user, the more likely he or she is to have concrete ideas about how to innovate—and the more important it is for higher-ups to listen.

Even a classic top-down innovator like Steve Jobs knew that he needed to listen to front-line workers because they might have figured out something about his products that he hadn't yet recognized. Top-down attempts at innovation usually wind up stifling it if managers and other leaders are unwilling to listen to "the little people" and insist on a one-question, one-answer approach. Luckily, in our personal lives, a certain way of innovating isn't prescribed. We are free to go where curiosity takes us.

Myth #5: You Can't Force Innovation

It's very true that you can't tell others to start innovating, pronto, and expect much good to come of it. But you *can* create an environment that encourages and rewards curiosity and therefore promotes engagement and innovation. When Canadian Tire faced a sales slump, the retailer didn't start frantically discounting tires and tools to try to get men interested again. Instead, the retailer hit control-alt-delete, wiped out all its assumptions, then gave its people carte blanche in terms of indulging their curiosity: ask what you want, get the answers how you want. It's also possible to kick-start innovation by focusing on creating a sense of connection and ownership, as the military found when it began soliciting, then responding to, suggestions from personnel.

Myth #6: Change Is Always Good

Tell that to the product team that dreamed up New Coke. The funny thing about that epic failure was that the beverage itself actually tested well—people liked the stuff. What *didn't* fly was the implication that there was something wrong with Old Coke. Mutiny ensued among consumers and the company wisely retreated, but equally wisely, continued to dream up new products, understanding that the occasional mistake is just part of the process of innovation. If the end goal remains clearly in sight, missteps become learning experiences, things not to repeat, and can actually wind up strengthening an organization. The sheer math on ideas suggests that about half of them will be lousy. But that's not catastrophic unless the lesson taken is that there's no point in continuing to dream up anything new, and it's safer to stick with what's always worked in the past.

Myth #7: Innovation Isn't for Everyone

Let's put this one to rest forever. Remember how kids in developing countries respond to the Soccket? When they see the ball, they almost immediately start asking questions and dreaming up their own innovations. Innovative thinking is contagious. It's a bug that anyone can catch.

Since our ancestors first stood upright, human beings have been innovating: more and better tools, different and improved circumstances, more effective and efficient ways of doing things. It's pretty silly to think we've suddenly all lost that basic drive now that we've hit the twenty-first century. If anything, our capacity to innovate is now exponentially greater because of our unprecedented ability to share information and ideas, which also makes it much easier to take something from one field and apply it to another.

INNOVATE FROM WITHIN

At heart, innovation is about approaching the world differently. It's about asking "Why?" and "Why not?"—questions that tripped off our tongues literally dozens of times an hour when we were little. But although curiosity may have been trained out of us, we can train it right back in.

One recent experiment illustrates how easy this can be. Researchers divided undergraduates into two groups and gave the first group a writing assignment: "You are seven years old. School is cancelled and you have the entire day to yourself. What would you do? Where would you go? Who would you see?" The second group got the same assignment, minus the first sentence. In other words, they remained adults (if undergrads can really be considered adults) in their minds. Next, both groups took a classic test to measure innovative, divergent thinking. Those who'd pretended they were little kids produced much more original responses, an effect that was particularly pronounced "among more introverted individuals, who are typically less spontaneous and more inhibited in their daily lives."

Apparently, simply thinking like a child, with all the benefit of adult analytical reasoning and knowledge, can help you think more innovatively. The authors noted that in business settings, simply mandating innovative thinking is unlikely to work because "doing so is quite antithetical to the manner in which curiosity and innovation naturally work." However, they continued, introducing a spirit of fun and play—anything that gets workers to think a little more "childishly"—might do the trick.

So how can we think more childishly? What does that even mean? It means asking questions without worrying about sounding foolish. It means listening without biases and preconceived notions. It means not rushing to solutions but really playing with problems, giving your brain time to consider all the options. It means persisting when you can't get the answer right away, and trying to think of another question that might get you there. It means accepting mistakes as non-successes, not shameful disasters. It means being kind to yourself when you blow it, so that you don't become paralyzed by fear of failure. It means throwing yourself into things and fully engaging, in the way that a three-year-old's mind—and indeed, entire body—seems alert to learning. And perhaps most importantly, it means knowing instinctively that questions and learning aren't just useful—they're fun. Curiosity can feel like your brain dancing joyfully.

Individual innovators and innovative businesses already know this, as we've seen. They are passionately curious, but often their questions are pretty basic: Who am I? What do others need from me? Where do I want to end up? How can I get there? Has someone else in another field already figured out a solution I can borrow? There's always more than one answer to such questions, of course. Sometimes there are dozens. So there's a follow-up question that helps you drill down to make choices: Why? Why x and not y? Hey, what about z?

These are questions all of us should be asking. We can use the approaches developed by businesses and by innovative thinkers to help turbocharge our curiosity. This will be good for our employers because it sets us up to be more innovative. And that will be good for Canada because innovative thinking is what's needed to address our productivity problem.

And it will also be good for *us.* Asking these questions can make our lives a whole lot better. At work, the moment we stop questioning and trying to improve is the moment we stop having fun. We check our passion at the door of the office, and work becomes a dreary shadow of what we really care about. At home, accepting the status quo makes us feel powerless to fix what's wrong. But we aren't. Asking questions that lead even to small changes can make us feel more engaged and more fulfilled. The trick is to be willing to embrace the answers, no matter where they take us.

Here's the best part about thinking like a little kid: you're *not* a little kid. You are able to think about how you think. Self-awareness is the adult trait that elevates curiosity to a new place, where it's not just fun but powerful because it fuels not only engagement and interest, but also actual, implementable innovation. In ways big and small, asking questions makes life richer, more interesting, more fulfilling and more complete. Better. That's the power, and ultimately the purpose, of *why*.

NOTES

The backbone of this book is based on a series of interviews I did in 2011 and 2012 with various types of innovators—dozens and dozens of hours spent with experts and specialists who were unfailingly gracious. They are, variously, inventors, big thinkers, creative types, and the fiercely curious—and all had a passion for whatever it is they do and a willingness to share it.

INTRODUCTION

Alessandro Di Fiore, the CEO of the European Centre for Strategic Innovation, cites the centre's study in his *Harvard Business Review* blog of March 19, 2012, at http://blogs.hbr.org/cs/2012/03/creativity_with_a_small_c.html.

Much has been written on the nature-versus-nurture debate as it relates to creativity. The seminal work in this area comes from the ongoing longitudinal Minnesota Twin Family Study, which includes research on twins separated at birth and raised in different families. See N.G. Waller, T.J. Bouchard Jr., D.T. Lykken, A. Tellegen and D. Blacker, 1993, "Creativity, Heritability, and Familiarity: Which Word Does Not Belong?" *Psychological Inquiry* 4: 235–37. Other research indicating that nurture trumps nature when it comes to creative abilities include F. Barron, *Artists in the Making* (New York: Seminar Press, 1972); S.G. Vandenberg, ed., *Progress in Human Behavior Genetics* (Baltimore: Johns Hopkins Press, 1968); R.C. Nichols, 1978, "Twin Studies of Ability, Personality, and Interests," *Homo* 29: 158–73; and R. Keith Sawyer, *Explaining Creativity: The Science of Human Innovation*, 2nd ed. (New York: Oxford University Press, 2012).

Dr. Ronald Beghetto at the University of Oregon has done a lot of interesting research on big-C versus little-c creativity. For a good recent overview, see R.A. Beghetto and J.C Kaufman, 2007, "Toward a Broader Conception of Creativity: A Case for 'Mini-C' Creativity," *Psychology of Aesthetics, Creativity, and the Arts* 1, 2: 73–79.

Although Federal Express grew through a series of strategic acquisitions over time, its genesis was a paper written by founder Fred Smith while he was a student at Yale in the late 1960s. Noting the logistical problems facing manufacturers as mass-produced electronics became more popular, he theorized that the only delivery system that would work would be by air. His idea, articulated in the college paper, was to have one carrier operate its own aircraft, depots, warehouses and vans. A single central hub would process all parcels. The paper was graded C, but Smith persevered with the idea, founding Federal Express Corporation in 1971. Beginning with just twenty planes, the business almost went under in its early going, but Smith again stuck to his guns—and on a trip home from Chicago, where he had been turned down by yet another investor, impulsively flew to Las Vegas and won $27,000, an amount that allowed FedEx to meet its payroll the following Monday. Though it was losing a million a month at the time, Smith saw the winnings as an omen and determined to stick with the business. Omen or not, it wasn't long before a consortium of venture capitalists put between $50 and $70 million into the start-up, making FedEx one of the most heavily financed U.S. companies in history at that time. Today FedEx is seen as a proxy for the U.S. economy, and thanks to the advent of online shopping, continues to thrive.

Alison Gopnik is a prolific researcher who has written two mainstream books explaining her fascinating research on how children think. My favourite is *The Scientist in the Crib: What Early Learning Tells Us about the Mind,* which she co-authored along with Andrew N. Meltzoff and Patricia K. Kuhl (New York: William Morrow, 1999).

CHAPTER ONE

The study on why kids ask why and how caregivers respond was jointly conducted research between the Universities of Michigan and Hawaii. I interviewed co-author Susan Gelman at the University of Michigan. An abstract of the research can be found at http://onlinelibrary.wiley.com/doi/10.1111/j.1467-8624.2009.01356.x/abstract, and the original article, here: Child Development 80, 6: 1592–611.

The study of 1,795 three-year-olds hypothesized that the reason curious kids become more intelligent ones by the age of eleven is that they "create for themselves an enriched environment that stimulates cognitive development." A. Raine, C. Reynolds, P.H. Venables and S.A. Mednick, 2002, "Stimulation Seeking and Intelligence: A Prospective Longitudinal Study," *Journal of Personality and Social Psychology* 82: 663–74.

Innovative Intelligence: The Art and Practice of Leading Sustainable Innovation in Your Organization, by David Weiss and Claude Legrand, is among the many books that have been written for businesses interested in fostering innovation, but it stands out for its attention to the individual. I particularly like the focus on the way our brains process information and the necessity of transferring information to the so-called limbic brain. Through his consulting firm, Ideaction, Legrand spends his time helping individuals learn to tap their innovative intelligence by learning to focus more on the question than the answer. But Legrand was also my first step into this world. It was his introduction that led me to many of the experts in this book, from Rolf Smith to

Matt Feaver, and I am extremely grateful. Like many of the people he connected me to, Legrand feeds off new ideas—it's the only currency that seems to really matter.

A misplaced finger or hand is the most common cause of table-saw accidents, and as Gass notes, human reflexes are slow enough that by the time the victim feels any pain he or she may have lost two fingers. The U.S. Consumer Product Safety Commission began, at SawStop's request, to review a proposal that would require all table saws to be equipped with a safety device that would stop on contact with flesh. The commission's first review was in October 2011, and it continues (at time of writing) to extend the period for submission. http://www.cpsc.gov/library/foia/foia11/brief/tablesaw.pdf.

The major manufacturers have so far resisted adopting the technology, despite at least one lawsuit based on their decision *not* to equip their saws with it.

"Intellectual hide-and-seek" is a term coined by Ronald Beghetto. R.A. Beghetto, 2010, "Prospective Teachers' Prior Experiences with Creativity Suppression," *International Journal of Creativity and Problem Solving* 20: 29–36.

A terrific thumbnail sketch of *The Innovator's DNA*, which explains the five major characteristics of innovators, was published in the December 2009 issue of *Harvard Business Review* and is available at http://hbr.org/2009/12/the-innovators-dna/sb2. The quotations I cite from Michael Dell and Ratan Tata are both included in that summary, on page 5.

Andrew Cosslett was interviewed by Adam Bryant for his Corner Office column on April 3, 2010.

http://www.nytimes.com/2010/04/04/business/04corner .html?_r=1&pagewanted=all.

TED is an acronym for Technology, Entertainment, Design, and it is a series of conferences globally owned by the Sapling Foundation. It began as a one-off conference in 1984 but has since become more frequent, with many offshoots. Speakers are given a maximum of eighteen minutes to present their ideas as creatively as possible. All TED talks are archived at www.ted.com, including Ken Robinson's, which has been viewed more than 10,000,000 times: http:// www.ted.com/talks/lang/en/ken_robinson_says_schools _kill_creativity.html.

In the report Robinson chaired for the UK government (http://sirkenrobinson.com/skr/pdf/allourfutures.pdf), he defined creativity this way: "Imaginative activity fashioned so as to produce outcomes that are both original and of value." Many innovation practitioners would recognize this as a good definition of innovation. The referenced longitudinal study on children and creativity is research documented in George Land and Beth Jarman's *Breakpoint and Beyond: Mastering the Future—Today* (New York: HarperCollins, 1992).

On why teachers who didn't like going to school are more likely to value creativity, see R.A. Beghetto, 2006, "Creative Justice? The Relationship between Prospective Teachers' Prior Schooling Experiences and Perceived Importance of Promoting Student Creativity," *Journal of Creative Behavior* 40, 149–62.

For more about student teachers' faith in rote learning and why they believe it should start in Grade 1, see R.A. Beghetto, 2008, "Prospective Teachers' Beliefs about Imaginative Thinking in K–12 Schooling," *Thinking Skills and Creativity* 3: 134–42.

Two eye-opening studies on teachers' dislike of creative students provide a very helpful summary of past research that confirms the same bias. Erik L. Westby and V.L. Dawson, 1995, "Creativity: Asset or Burden in the Classroom?" *Creativity Research Journal* 8, 10: 1–10.

Alison Gopnik's long Q&A was published in the November 1, 2011, edition of *Maclean's*.

The worker output per hour measurement comes from a Conference Board of Canada report on our productivity: http://www.conferenceboard.ca/hcp/details/economy/measuring-productivity-canada.aspx. The Conference Board notes: "Low productivity levels present an enormous challenge for Canada's future economic prosperity. In 2008, Canada's level of productivity was US$35, much lower than that of the United States, at US$44. This earned Canada a disappointing 16th place among its 17 peer countries on the level of labour productivity. . . . Despite a broad and growing consensus that Canadian productivity needs to be improved, the gap with the U.S. is widening, not narrowing." A separate Conference Board report noted the effect on corporate profit and personal wealth: http://www.conferenceboard.ca/press/newsrelease/11-11-16/two_decades_of_sluggish_productivity_growth_costly_to_canadian_businesses_governments_and_individuals.aspx.

The Conference Board also determined that government revenues could have been $66 billion higher had we matched the U.S. rate of productivity: http://www.conferenceboard.ca/e-Library/abstract.aspx?did=3396.

The study on how to promote divergent thinking in kindergarten makes it clear that this is not only simple to do but also something that parents as well as educators should already be doing. Mary Jo Puckett Cliatt, Jean M. Shaw and Jeanne M. Sherwood, 1980, "Effects of Training on the Divergent-Thinking Abilities of Kindergarten Children," *Child Development* 51, 4: 1061–64.

Alison Gopnik describes the MIT experiment in her interview with *Maclean's* and also discussed it in a TED talk she gave: http://www.ted.com/talks/lang/en/alison_gopnik_what_do_babies_think.html.

The experiment with science teachers is a good indication that adults as well as children benefit hugely from the opportunity for open-ended questioning. Robert E. Yager, Nor Hashidah Abd-Hamid and Hakan Akcay, 2005, "The Effects of Varied Inquiry Experiences on Teacher and Student Questions and Actions in STS Classrooms," *Bulletin of Science, Technology & Society* 25, 5: 426–34.

For a good review of the research on the link between curiosity and happiness, see Matthew W. Gallagher and Shane J. Lopez, 2007, "Curiosity and Well-Being," *Journal of Positive Psychology* 2, 4: 236–48. The quotation relating curiosity to stretching is from a hugely interesting study by leading researchers: Todd B. Kashdan, and Michael F. Steger, 2007, "Curiosity and Pathways to Well-Being and Meaning in Life: Traits, States, and Everyday Behaviors," *Motivation and Emotion* 31: 159–73, quotation at 171.

CHAPTER TWO

Canadian Tire's 2009 annual report can be found at http://corp.canadiantire.ca/EN/Investors/FinancialReports/Annual%20Reports%20Library/CTC_AR_2009.pdf, but the characterization of its performance was drawn from interviews with insiders at that time. You can read more about Wetmore's perspective here: http://www.theglobeandmail.com/report-on-business/rob-magazine/how-to-fix-canadian-tire/article584299/?page=2.

Roger Martin has written extensively about innovative thinking. The reference to relativity bias is from *The Design of Business,* 37, 42–43. But this and other Martin citations are also drawn from lengthy interviews with him.

CHAPTER THREE

One of the most fascinating discoveries for me was how the military is extremely innovative. As Legrand notes in *Innovative Intelligence* and I discovered, some of the top-ranking innovators are former military. Rolf Smith is a good example. His *The 7 Levels of Change* is aimed squarely at helping people change their mindsets about being innovators.

CHAPTER FOUR

I've taken some licence with this story to make it more palatable to anyone who doesn't work on Bay Street. The very first bought deal was actually before the Bank of Nova Scotia transaction, a $200-million perpetual preference share deal that Gordon Capital did with RBC in 1982. It was that deal that caused Connacher to say to Eberts, "I hope you know what the fuck you are talking about because nobody else does." And it was also that deal that forced every investment bank in the world to begin doing bought deals.

CHAPTER FIVE

If Addison Lawrence makes you feel like becoming a shrimp farmer—I have to admit I toyed with the idea after talking to him—you can find out more about the economics of the business from a white paper by Royal Caridea, which purchased the rights to his stacked raceway methodology. This is where I got the U.S. consumption and importation figures, which Dr. Lawrence also confirmed. http://www.nist.gov/tip/wp/pswp/upload/188_sustainable_shrimp_aquaculture_alternative.pdf.

Peter Löscher, CEO of Siemens, was interviewed by Adam Bryant for his Corner Office column on July 30, 2011. http://www.nytimes.com/2011/07/31/business/siemens-ceo-on-building-trust-and-teamwork.html.

Dan Satterthwaite, head of human resources at DreamWorks Animation, was quoted in "How Three Companies Innovate," *New York Times,* March 17, 2012. http://www.nytimes.com/2012/03/18/business/how-three-companies-innovate.html.

Regina Dugan's TED talk can be viewed at http://www.ted.com/talks/lang/en/regina_dugan_from_mach_20_glider_to_humming_bird_drone.html. Emphasis in quotation is mine.

Harvard prof Amy Edmondson wrote about failure—and the need to learn from it—in the *Harvard Business Review*, http://hbr.org/2011/04/strategies-for-learning-from-failure/ar/1.

For more on how a little kindness to yourself can go a long way in terms of promoting innovative thinking, see Darya L. Zabelina and Michael D. Robinson, 2010, "Don't Be So Hard on Yourself: Self-Compassion Facilitates Creative Originality among Self-Judgmental Individuals," *Creativity Research Journal* 22, 3: 288–93.

CHAPTER SIX

The best source of information on all things TRIZ is the website www.triz-journal.com, where many articles on various applications of TRIZ are archived. The site includes a clear layperson's introduction to TRIZ by Katie Barry, Ellen Domb and Michael S. Slocum: http://www.triz-journal.com/archives/what_is_triz/. For a lively journalistic account of Altschuller's life, see Mark Wallace's article "The Science of Invention" in Salon.com, at www1.salon.com/tech/feature/2000/06/29/altshuller/index.html.

Though I interviewed him directly, Isadore Sharp's reflections can also be found in his book *Four Seasons: The Story of a Business Philosophy*.

That the merger of AOL and Time Warner was a failure is widely acknowledged, if only because of the collapse in shareholder value. Here's what Time Warner CEO Jeffrey Bewkes had to say: http://www.telegraph.co.uk/finance/newsbysector/mediatechnologyandtelecoms/media/8031227/AOL-merger-was-the-biggest-mistake-in-corporate-history-believes-Time-Warner-chief-Jeff-Bewkes.html. Some of the direct references to the deal coming together are pulled from this *New York Times* story about the transaction: http://www.nytimes.com/2010/01/11/business/media/11merger.html?_r=1&pagewanted=all.

CHAPTER EIGHT

Some of Gary Hamel's references are drawn from a speech I attended in 2011, but they are backed up by his book *The Future of Management*, 57–58.

The statistics on handwashing by health care providers relies on data gathered by the Ontario Ministry of Health in 2010, which showed that health care providers were washing their hands at a rate of 67 percent as they entered a patient's room. The 70 percent figure I used is sourced from a senior hospital administrator in 2012.

Holly White and Priti Shah have been researching the connection between divergent thinking and ADHD and have authored several studies showing that the same factors that make it difficult for people with ADHD to ignore distractions also make what White and Shah call "the collision of ideas" much easier. See Holly A. White and Priti Shah, 2006, "Uninhibited Imaginations: Creativity in Adults with Attention-Deficit/Hyperactivity Disorder," *Journal of Personality and Individual Differences* 40: 1121–31; Holly A. White and Priti Shah, 2011, "Creative Style and Achievement in Adults with Attention-Deficit/Hyperactivity Disorder," *Journal of Personality and Individual Differences* 50, 5: 673–77.

CHAPTER NINE

One of the best overviews of the scholarly research on brainstorming can be found in "Beyond Productivity Loss in Brainstorming Groups: The Evolution of a Question," a chapter in an academic text that is nevertheless highly accessible. W. Stroebe, B.A. Nijstad and E.F. Rietzschel, in *Advance in Experimental Social Psychology,* vol. 43, ed. M.P. Zanna and J.M. Olson (Burlington, MA: Academic Press, 2010), 157–203. The authors, who are leading researchers on brainstorming, review decades of research demonstrating that groups that are comfortable with one another tend to be less creative, and explaining why turn-taking blocks the production of ideas. The quotation I cite regarding the ways that turn-taking interferes with idea generation—by disrupting image activation and interrupting a train of thought—is from page 174. For the classic journal article on the idea of production blocking, see Michael Diehl and Wolfgang Stroebe, 1987, "Productivity Loss in Idea Generating Groups: Toward a Solution of the Riddle," *Journal of Personality and Social Psychology* 53: 497–509. Studies on the efficacy of electronic brainstorming (and brainstorming via pen and paper) include R.B. Gallupe, L.M. Bastianutti and W.H. Cooper, 1991, "Unblocking Brainstorms," *Journal of Applied Psychology* 76: 137–42, and R.B. Gallupe, W.H. Cooper, M. Grise and L.M. Bastianutti, 1994, "Blocking Electronic Brainstorms," *Journal of Applied Psychology* 79: 77–86.

CHAPTER TEN

The housing-free elevator struck my imagination because as a design student I felt the housings always ruined the tops of buildings—however elegantly designed, the necessity to add workings for elevators marred the roofline. I interviewed Pedro Baranda at length about his experience helping to solve the problem. He has since been promoted to president of Otis Elevator Company, a subsidiary of Otis Worldwide.

Many studies demonstrate the connection between innovative thinking and having lived abroad, but one of the best, because it tests five hypotheses and also explains why travelling abroad doesn't similarly affect creativity, is quite recent. William W. Maddux and Adam D. Galinsky, 2010, "Cultural Borders and Mental Barriers: The Relationship between Living Abroad and Creativity," *Journal of Personality and Social Psychology* 96, 5: 1047–61.

For more on construal level theory, see Yaacov Trope and Nira Liberman, 2010, "Construal-Level Theory of Psychological Distance," *Psychological Review* 117, 2: 440–63. For research on how distance, even hypothetical distance, improves problem-solving abilities, see L. Jia, E.R. Hirt and S.C. Karpen, 2009, "Lessons from a Faraway Land: The Effect of Spatial Distance on Creative Cognition," *Journal of Experimental Social Psychology* 45: 1127–31. For research on how problem solving is easier to do for others, including several variations of the prisoner-in-a-tower puzzle, see Evan Polman and Kyle J. Emich, 2011, "Decisions for Others Are More Creative than Decisions for the Self," *Personality and Social Psychology Bulletin* 37, 4: 492–501. The first reference to the prisoner in the tower riddle may have been made in Janet Metcalfe and David Wiebe, 1987, "Intuition in Insight and Noninsight Problem Solving," *Psychonomic Society Inc*, 238.

Nick Dalley's company is called Intentional Communication Inc., and is based in Dripping Springs, Texas.

Rosenfeld's ISPI is a fascinating personality test tool that I asked members of my family to take. Knowing the ISPI "totem" of people close to you is a useful way of considering how you interact and why. As I was researching the book I also spoke to a group of senior psychologists who had been trained to administer the ISPI and plan to use it in their practices.

CHAPTER ELEVEN

Dick Brass's op-ed titled "Microsoft's Creative Destruction" appeared in the *New York Times* on February 4, 2010, http://

www.nytimes.com/2010/02/04/opinion/04brass.html.

Rick Wagoner's statement to Congress was quoted in "At G.M., Innovation Sacrificed to Profits," an article by Micheline Maynard in the *New York Times,* December 5, 2008, http://www.nytimes.com/2008/12/06/business/06motors.html.

The observations about Ford's quarter are mine (and those of analysts who follow the company), but my assumptions about the strategy behind it are entirely conjecture, based on its success with increasing the margin per vehicle of more fuel-efficient vehicles. You can read the company's own report on the quarter at http://media.ford.com/article_display.cfm?article_id=34463.

For additional reading on Tata's Nano, see http://finance.ninemsn.com.au/executivesuite/motoring/8399162/indias-tata-admits-mistakes-with-nano.

CONCLUSION

The study on how pretending to be a seven-year-old stimulates innovative thinking also provides a useful overview of research on creative thinking. Darya L. Zabelina and Michael D. Robinson, 2010, "Child's Play: Facilitating the Originality of Creative Output by a Priming Manipulation," *Psychology of Aesthetics, Creativity, and the Arts* 4, 1: 57–65.

ACKNOWLEDGEMENTS

This book could not have been written without the vast knowledge and expertise of the group of people I mentally call "the innovators"—pioneers of thought who blazed trails and generously shared their insights with me. They are Roger Martin, Rolf Smith, Robert Rosenfeld, Andrew Harrison, Matt Feaver, Jana Meerkamper, Cedric Paivin, Rob LaJoie, Chris Stamper, Peter Engstrom, Alexander Manu, Barry Jaruzelski and, of course, one of the first great thinkers on the subject, Clayton Christensen. This book also contains insight and wisdom from Kevin Lynch, Kathy Bardswick, John Ruffolo, Jack Diamond, Mark Aboud, Les Dakens and Bonnie Brooks. A very special thanks to Claude Legrand, without whom I would not have met many of these innovators and who inspired me with his wisdom on and passion for the subject.

The stories in this book are about extraordinary people

maximizing their potential, and I am brimming with gratitude for their generosity in sharing their experiences with me: Addison Lawrence, Steve Gass, Sean Moore, Julia Silverman, Isadore Sharp, Pedro Baranda, Jean Blacklock, Michael Budman, Maria Valente, Brian Arthur, David Helfand, Maymie Tegart and, last but never least, Gordon Eberts. Their stories—from the saw that will save your finger to the soccer ball that can power a light and also a child's imagination—are the most important part of this book, the part that brought me the most joy. I believe passionately that the rest of us can learn from their examples, and if nothing else, learn to enjoy what we do and do it with the same enthusiasm.

Every book needs first readers, and so to Adrian Lang, Corinne Pruzanski and Maya Carvalho, thanks beyond words for being that for me. And for being gentle. Thanks to Kate Fillion for your support and advice, worth its weight in gold. There aren't thanks enough for your great friendship. Thanks to Matt Fairley. To Rick Broadhead, who persuaded me that I could in fact find the time to write a book and whom I only cursed occasionally, this really wouldn't have happened without you. To Brad Wilson at HarperCollins, thank you for making it a better book. Thanks to the whole team at HarperCollins for your enthusiasm and support. Enormous thanks to my family, for putting up with me as this got written—and loving me anyway. And finally, Vince Borg knows that this project would not have been started, let alone finished, without his support and encouragement. For that—and so much else—he has my eternal gratitude.

INDEX